Datums and Map Projections

Datums and Map Projections

for Remote Sensing, GIS and Surveying

2nd Edition

Jonathan Iliffe
Roger Lott

Whittles Publishing

Published by
Whittles Publishing,
Dunbeath,
Caithness KW6 6EG,
Scotland, UK
www.whittlespublishing.com

© 2008 Jonathan Iliffe & Roger Lott
2012, 2014 reprinted with corrections
ISBN 978-1-904445-47-0

First edition published 2000.

Every effort has been made to trace copyright
holders and to obtain their permission for
the use of copyright material. The publisher would
be grateful if notified of any amendments that
should be incorporated in future reprints or
editions of this book.

Printed and bound in Great Britain
by Severn, Gloucester

CONTENTS

Preface   ix

Chapter 1: Introduction   1
  1.1   The context   1
  1.2   Introduction to the concepts and the structure of this book   2

Chapter 2: Coordinates and reference systems   8
  2.1   The Earth – geoid and ellipsoid   8
      2.1.1   The geoid   8
      2.1.2   Models of the shape of the Earth – the ellipsoid   8
  2.2   Coordinate systems   11
      2.2.1   Coordinate system attributes   11
      2.2.2   Coordinate systems for the sphere and ellipsoid   12
      2.2.3   Geocentric Cartesian coordinates   14
      2.2.4   Conversion between ellipsoidal and geocentric
            Cartesian coordinates   15
      2.2.5   Map projection coordinates   16
      2.2.6   Cartesian coordinates for engineering applications   17
      2.2.7   Gravity-related systems (height and depth)   18
      2.2.8   Miscellaneous coordinate systems   19
  2.3   Datums and coordinate reference systems   20
      2.3.1   Datum overview and classification   20
      2.3.2   Geodetic datums and coordinate reference systems   20
      2.3.3   Projected coordinate reference systems   28
      2.3.4   Vertical systems   29
      2.3.5   Engineering datums and coordinate reference systems   35
      2.3.6   Image datums and coordinate reference systems   35
  2.4   Compound coordinate reference systems   36
  2.5   Coordinate reference system identification   36
      2.5.1   CRS description   36
      2.5.2   Registers of coordinate reference systems   38

Chapter 3: Map Projections   39
  3.1   Introduction   39
  3.2   Map projections: fundamental concepts   40
      3.2.1   Grids and graticules   40
      3.2.2   Scale factor   40
      3.2.3   Developable surfaces   42
      3.2.4   Preserved features   44
      3.2.5   Spheres and ellipsoids   46

3.3	Cylindrical projections	48
	3.3.1 Cylindrical equidistant	48
	3.3.2 Cylindrical equal area	51
	3.3.3 The Mercator projection	53
	3.3.4 Transverse Mercator	55
	3.3.5 South-oriented Transverse Mercator	61
	3.3.6 Oblique Mercator	61
3.4	Azimuthal projections	63
	3.4.1 General azimuthal	63
	3.4.2 Azimuthal equidistant	63
	3.4.3 Azimuthal equal area	64
	3.4.4 Stereographic	66
	3.4.5 Gnomonic	69
	3.4.6 Azimuthal orthographic	70
	3.4.7 Azimuthal perspective projection	70
3.5	Conic projections	73
	3.5.1 General conic	73
	3.5.2 Conic equidistant	75
	3.5.3 Albers equal area	75
	3.5.4 Lambert Conformal Conic	77
	3.5.5 Oblique conic	79
3.6	Non-geometric projection methods	80
3.7	Summary of information required	83
	3.7.1 Map projection method formulae	83
	3.7.2 Map projection parameter values	84
3.8	Computations within map projections	86
3.9	Designing a map projection	89
Chapter 4: Transformations		**90**
4.1	Introduction	90
4.2	General characteristics of transformations	91
	4.2.1 Transformations and conversions	91
	4.2.2 Transformation multiplicity	91
	4.2.3 Transformation accuracy	92
	4.2.4 Transformation reversibility	93
4.3	Transformations between geocentric coordinate reference systems	93
	4.3.1 Introduction	93
	4.3.2 Three parameter geocentric transformation	94
	4.3.3 Seven parameter geocentric transformation	95
	4.3.4 Ten parameter geocentric transformation	98
4.4	Transformations between geographic coordinate reference systems	100
	4.4.1 Introduction	101
	4.4.2 Molodensky and Abridged Molodensky	101
	4.4.3 Geographic offsets	102
	4.4.4 Grid interpolation – NTv2 and NADCON	104
	4.4.5 Indirect transformation between geographic coordinate reference systems	106

4.5	Transformation of 2D plane coordinates	109
	4.5.1 Introduction	109
	4.5.2 Compatibility of coordinate reference systems	110
	4.5.3 Similarity transformation method	113
	4.5.4 Affine transformation	115
	4.5.5 Polynomials	117
	4.5.6 Creating overlays in Google Earth™	118
	4.5.7 Transformation of GPS data onto a local site grid	120
	4.5.8 Indirect transformations between projected coordinates	122
4.6	Coordinate operations for vertical coordinate reference systems	123
	4.6.1 Introduction	123
	4.6.2 Vertical offsets	124
	4.6.3 The hub concept	127
4.7	Transformation between ellipsoidal and gravity-related heights	128
	4.7.1 Geoid models	128
	4.7.2 Height correction models	129
	4.7.3 Transformations involving compound coordinate reference systems (CRSs)	130
4.8	Selecting a transformation	131
	4.8.1 Introduction	131
	4.8.2 Officially sanctioned transformations	132
	4.8.3 Selecting from a transformation repository	132
4.9	Deriving your own transformation	134
	4.9.1 Introduction	134
	4.9.2 Choice of transformation method	134
	4.9.3 Availability of control points	135
	4.9.4 Geometric issues	136
	4.9.5 Effect of ignoring geoid-ellipsoid separation	137
	4.9.6 Evaluating results of the transformation	140

Chapter 5: Global Navigation Satellite Systems — **141**
5.1	Introduction	141
5.2	The systems	141
5.3	Positioning with codes	143
5.4	Differential GNSS and augmentation systems	146
5.5	GNSS measurements using phase observations	148
5.6	Coordinate reference system considerations	152

Chapter 6: Case Studies — **154**
6.1	Transformation of GPS data into a local coordinate reference system	154
6.2	Creation of a three-parameter geocentric transformation from an official national transformation	160
6.3	Designing a map projection	162
6.4	Calculations using map grid coordinates	164
6.5	Creating overlays in Google Earth™	172

Appendix A: Terminology   **176**

Appendix B: Computations with spherical coordinates   **181**

Appendix C: Basic geometry of the ellipsoid   **182**
    C.1   Introduction   182
    C.2   Radii of curvature of the ellipsoid   182
    C.3   Normal sections and geodesics   182
    C.4   Forward computation of coordinates   184
    C.5   Reverse computation of azimuth   184
    C.6   Determination of points on the geodesic   185

Appendix D: The Molodensky equations   **186**

**Appendix E: Determination of transformation parameter values
         by least squares**   **187**
    E.1   Introduction and least squares terminology   187
    E.2   Two dimensional transformations of Cartesian coordinates   189
           E.2.1   The Similarity transformation   189
           E.2.2   The affine transformation   192
           E.2.3   Second order polynomials   193
    E.3   Three-dimensional transformations of Cartesian coordinates   194
           E.3.1   The seven-parameter transformation   194
           E.3.2   The ten-parameter geocentric transformation   196
           E.3.3   Subsets of the seven-parameter geocentric transformation   196
    E.4   Worked example   197

References & Further Reading   **200**

Index   **203**

PREFACE

It is now eight years since the first edition of *Datums and Map Projections* was originally published. In that time the book has found its way onto the shelves of many students and professionals around the world, so it seems to be fulfilling a need.

The theme of the book – a practical guide to coordinate reference systems – is as important now as when it was first published – probably more so when we consider the ever growing use of satellite navigation systems and the introduction of web mapping services such as Google Earth™. A second edition seems timely, and here it is.

The first change that a reader will notice is the authorship. The author of the first edition, Jonathan Iliffe, was delighted when Roger Lott, formerly chief surveyor of BP, agreed to share the task of editing, re-writing, revising, updating and expanding the original text. Between us we have made some very substantial changes.

In the first place we have changed the structure of the book to one that we believe gives a better grouping of common themes. Secondly, we have enlarged the scope so that the book covers all possible different types of coordinate reference system that are used in mapping and related areas, from ellipsoidal latitude and longitude through to depths below chart datum. Thirdly, since we now clearly recognise that the book has an international market, we have brought in many more examples and case studies from around the world. Fourthly, the book has adopted the terminology of the ISO 19111 standard.

These are the main changes. To these can be added the usual editing and re-thinks about the way that the material was written, and the addition of colour diagrams.

We should like to place on record our gratitude to many colleagues who have taken the time to discuss different aspects of the book with us, including Paul Cross, Ian Dowman, Jeremy Morley, and Noel Zinn. We should also like to thank Paul Batey, Roel Nicolai, Anna Bakare, and Ali Al-Shaery for help with preparing diagrams and providing data for examples.

The various sections of the book have been back and forth between us many times and we can hardly remember who wrote which sentence. Therefore we jointly accept all responsibility for any remaining errors.

Jonathan Iliffe
Roger Lott

1

INTRODUCTION

1.1 The context

This book is designed for those dealing with spatially referenced data as a practical guide to the problems that may be encountered with datums and map projections. It is aimed at those working in surveying, remote sensing, geographic information systems, and related areas, and therefore covers a wide range of scales and accuracy targets. It is also likely to be the case that those encountering these topics will have very different starting points in terms of their level of knowledge: those who are aware that they need to know something about the subject, but are not yet sure what, would do well to commence by reading section 1.2 before proceeding further.

Until recently, an in-depth knowledge of datums was generally confined to a fairly small group of scientists and to geodesists working in government survey departments. Most surveying was carried out on local scales using terrestrial equipment (such as theodolites, levels, and distance measurers) to establish positions with respect to nearby control points: although the map projection was usually a consideration, the question of the datum was of minimal importance. For the general user requiring access to spatial data (on land use, for example), almost the only practical source was a published map.

Two developments in recent years have been most responsible for changing this state of affairs. One has been the explosion in the use of spatial data that has been brought about by the development of geographic information systems (GIS) for handling and manipulating data in digital form. The other has been the development of techniques such as the global positioning system (GPS), and satellite (or airborne) remote sensing, which have made available entirely new methods of acquiring accurate data. Moreover, these are techniques that, due to their global, space-based nature, have become completely divorced from the localised survey. In short, this is a classic case of supply and demand: more and more data are available from a variety of different sources, and more and more people have the means and the need to make use of them.

Supply, demand, and – potentially – confusion. A situation now exists where it is quite common for the means of acquiring data to be using a reference system that is completely different to the one in which the data will ultimately be required. This is a particular case of the problem of combining data sets, in which new data are to be put alongside archive material (which could have been referenced to one of many scores of possible datums and an uncountable number of possible map projections), but the problem is a more general one. A few examples will serve to illustrate this.

- Geo-referencing a satellite image with ground control points that have been established using GPS, and with others that have been obtained from a published map.

- Combining digital map data from two different survey organisations, for example as part of a cross-border collaboration between neighbouring states.
- Carrying out a survey with high precision GPS, and bringing it into sympathy with existing mapping in a local coordinate reference system.
- Navigating a ship or an aircraft using satellite systems, on charts that have been prepared in a national system. Even a leisure user of hand-held GPS receivers will need to make an appropriate correction when using them in conjunction with a map.
- Making an overlay in Google Earth™ from a digital map that uses a different projection and datum from the web based mapping.

These examples are significant in that they can each be said to represent an increasing trend. Indeed, it could be argued that the same forces that drive *globalisation* in the political and economic spheres are having their effect on trends in the area of spatial information. After all, a hydrographic surveyor in the English Channel and a land surveyor in Ghana could both simultaneously be using the same positioning satellite. They may not, at present, be expressing the final result in the same datum and reference system, but there is certainly a trend towards the adoption of regional and global datums.

The examples given above focused on the problems of combining data from different sources. Another reason for understanding the nature of a reference system is that it is often necessary to carry out computations using the data: it must, after all, have been acquired for some purpose. Computations that use geographic coordinates (latitude, longitude, and height) require a particular branch of mathematics. Map projections introduce distortions.

In some situations, this is hardly a problem. For example, if digital data from a national mapping organisation is being used to determine the optimum route on the road network between two points, the distortions of the projection are insignificant at the accuracy level required. To take an example at the other extreme of accuracy requirements, however, the establishment of maritime boundaries between states involves determining distances between median lines and coastal points. This computation can be done using geodetic coordinates, but not without difficulty. Using a projection is easier, but the distances will be distorted and the computation will be inexact: whilst this may be insignificant over small areas if an appropriate projection is used, it is not likely to be sufficiently accurate for large areas. Where the limit lies, and what an appropriate projection is, are questions it is necessary to be able to answer when tackling problems of this nature.

This book therefore aims to be a comprehensive and practical guide to all of these problems, accessible to users from a wide variety of backgrounds and with different levels of starting knowledge, and appropriate to any geographic region. We shall start by giving an introduction to some of the fundamental concepts in section 1.2, giving also an explanation of the structure of this book and where the different concepts are treated in more depth.

1.2 Introduction to the concepts and the structure of this book

Consider each of the following statements:

- The height of the point is 3.122 m.
- The height above mean sea level is 10.983 m.
- The latitude is 32°10'12.23" North; the longitude is 59°08'43.79" East.
- The coordinates of the point are 252 345.834, 204 301.227.

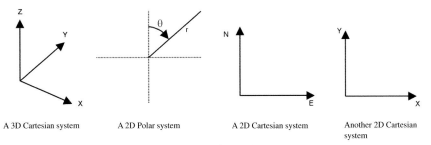

A 3D Cartesian system   A 2D Polar system   A 2D Cartesian system   Another 2D Cartesian system

Figure 1.1 **Examples of coordinate systems.**

All of these express coordinates[1] with a great deal of precision: the other thing that they have in common is that they are all – to a greater or lesser extent – ambiguous statements, as they contain no information on the coordinate framework in which they are measured.

To begin to define things a little more clearly, we must first start by understanding what we mean by a *coordinate system*. Some very simple examples of these are shown in Figure 1.1. The many different forms of coordinate system are explored more fully in chapter two, but we can see immediately that there are some fundamental questions that it is necessary to answer before someone using the coordinates can start to make sense of them. Is this a two-dimensional system or a three-dimensional one? Does it use distances along axes, or are angular measurements involved? What do we call the axes? In which order are they quoted? What units are used? As we shall see, the more complex the nature of the coordinate system, the more questions we shall have to answer in order to define exactly what we mean.

The coordinate system defines the set of axes that span the coordinate space and the number of axes will be the same as the dimension of the system. The coordinate system defines attributes of the axes – the direction in which coordinates increment along each axis, names, abbreviations and units for each axis, and, where the dimension is two or more, the order of the axes. The order of the axes determines the order in which coordinates are recorded. If any of these attributes are changed, the coordinate system has changed.

Once we have obtained this fundamental information about the coordinate system, we can begin to carry out some basic computations. We might be able to say how far apart two points are in an XYZ system if we know the coordinates of both of them. We might be able to say in what direction we need to go to get to one point from another if we know their coordinates in a two-dimensional system. However, what we cannot do yet is to describe the position of a particular point: to do this we need to know the relationship of the coordinate system to the particular object for which it is being used. In Figure 1.2, we see the *same coordinate system* being used in association with different objects.

In the examples shown in Figure 1.2, we might define the position of the origin in terms of a particular location on the object; we might define the directions of the axes with reference to the directions of the wings, or the fuselage, or fore/aft and port/starboard. Although the coordinate systems are the same (three-dimensional Cartesian, in the order XYZ, using metres as the units, and so on), we should certainly not describe the coordinates of a point on the aircraft in terms of the coordinate system attached to the ship – and we would certainly not try and do any computations that switched

---

[1] In some texts a single value is called an 'ordinate', with 'coordinates' implying multiple values. In this book 'coordinate' is used as the singular.

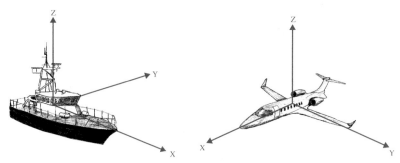

Figure 1.2 **Examples of coordinate systems referenced to different objects.**

casually from one system to the other. In fact what we have done is to introduce a *coordinate reference system* (CRS) – that is, a coordinate system that has been *referenced* to an object through describing the *datum*. In these cases we could think of the datum as being almost a physical mark on the ship or the aircraft that defined the origin, although clearly the orientation of the axes is as important a part of the referencing.

In this book, the object that we are most concerned with referencing coordinate systems onto is the Earth itself. What we need to understand is that there are many different ways – both in theory and in practice – of referencing any coordinate system, and this is what leads to the ambiguity. This is discussed in detail in Chapter 2. Coordinates are unambiguous only when they are clearly associated with a reference system and a clear definition of that system is available.

Let us start with a simple example of how a coordinate system can be referenced to the Earth in two or more different ways – in this example, a one-dimensional height system. If, for example, an engineer is interested only in the relative heights of points within a construction project site, and not in their relationship to the outside world, then it will be acceptable to designate one point as having an arbitrary height – 100 m for example – and find the heights of all others with respect to this. Such an act has established the position of the origin of the coordinate system with respect to the construction site and the orientation of the coordinate system is implied as being positive upwards.

This has defined the datum for the construction site vertical coordinate system. A datum defines the origin and orientation of a coordinate system with respect to an object.

It would also have been possible to define the datum by designating the height of another point, in which case all heights across the site would be different. Figure 1.3 illustrates a point with different heights in system A and system B.

If we have a height and we do not know whether it is in system A or in system B, we cannot relate it to the construction site. Thus, the identification of the datum is seen to be essential information that must accompany the coordinates. However, going further

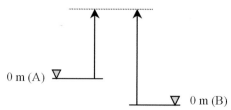

Figure 1.3 **An example of a point having different coordinates with respect to two different coordinate reference systems (in this case height only).**

than this, we must also consider the case where the engineer has not only established an initial starting point, but has also levelled from this to other control points around the site and therefore provided a framework of heights for use in the project. Another engineer establishing system B might establish slightly different relative heights between points, or may not use the same control points at all. Thus there may be a more complex relationship between the two systems than a simple offset, and this process of coordinate *realisation* means that 'the datum' may be not just an initial point and a coordinate direction, but a complete coordinate set.

What we shall see in Chapter 2 is that this concept also applies to two- and three-dimensional coordinate systems. Included in these are *ellipsoidal systems* that describe positions using *latitude* and *longitude*. We therefore have the important – and often not well understood – result that latitude and longitude are *not* unambiguous identifiers of a point's position. A point will usually have different latitude and longitude coordinates in a national system to what would be obtained through using a system such as GPS.

This then brings us on to the topic of *coordinate transformations* – that is, we have data in one reference system and we wish to match it onto another system. Several reasons for doing this, and examples of when there would be a disparity between systems, have already been introduced in section 1.1; many more are identified in later sections of this book. We could also relate this to the examples of referenced coordinate systems that we have already introduced: how does the XYZ system referenced to the ship in Figure 1.2, in which its echo sounder is located, relate to the coordinate reference system of the chart for which the survey is being carried out?

What we need to note in this introduction is that there are many different methods of coordinate transformation that can be identified. The first reason for this is that there are many different kinds of coordinate system, and so there need to be methods for transforming between systems of a similar type, as well as between systems of different types. Secondly, we are often dealing with cases where the data in one system is in some way distorted or inaccurate, or incomplete information exists about how its coordinate reference system was originally defined, or we are looking for a simple transformation rather than an accurate one. All these reasons mean that we are often presented with different *options* for how we transform from one coordinate reference system to another, and we need to understand the advantages and disadvantages of each technique.

There is one final concept that it is worth describing in this introduction, before we get into the detailed descriptions of later chapters, and this is the idea of *coordinate conversions*. This is where we go between two different *types* of coordinate system that have been referenced to an object *in the same way*, and it is thus distinct from a coordinate *transformation*. A simple example of this is illustrated in Figure 1.4, in which we illustrate two different methods of describing a point's position – Cartesian and polar. We can convert between the two, but we note that there is no change to the way in which they are referenced to the object: the *datum information* is the same for both of them. This is therefore a different *coordinate operation* to transforming to a new set of axes with a different origin and orientation.

An extremely important and common class of coordinate conversion is the *map projection*. This is where we convert from an ellipsoidal coordinate system to a two-dimensional Cartesian one. Another important type is the conversion from ellipsoidal systems to Earth-centred Cartesian systems. The latter conversion can be discussed in a few lines and the relevant formulae given, and has therefore been treated in this book at the point at which the coordinate system types are introduced in Chapter 2. Map projections, on the

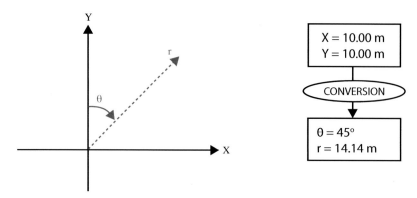

Figure 1.4 **An example of a coordinate conversion.**

other hand, form such a vast subject area that we have chosen to devote a whole separate chapter to them in this book (Chapter 3), after briefly introducing the topic in Chapter 2.

Another complicating factor in discussing coordinate conversions is that whilst on many occasions they are quite separate from coordinate transformations, there are situations in which the two are intrinsically bundled together as one operation. An example would be when trying to overlay a satellite image onto an underlying map of the same area: the geometry of the image and of the underlying map projection are different, and so there is a coordinate conversion involved, but there is also a coordinate transformation as there are different reference systems involved. We do not much mind about the difference – we just want to get from one system to the other and make the two data sets match. Therefore we should note that some of the material in chapter four on coordinate transformations applies also to coordinate conversions.

All of this information is summarised in Figure 1.5, which can be used as a 'route map' in navigating between different coordinates. Several different coordinate reference systems (CRSs) are shown, some on the same datum and some on different ones. *Conversions* (which do not involve a change of datum) are shown as dotted lines, and the paragraph numbers next to them show where in this book to find a description of the procedure. *Transformations* (which do involve a change of datum) are shown as dashed lines, and these are similarly labelled with the appropriate section numbers. So, for example, if the reader wishes to transform between two different vertical datums, it will be seen that CRS 9 and CRS 10 are two such examples, and the transformation is covered in section 4.6.

This figure is repeated at the start of Chapter 4 for ease of reference, and each description of a transformation highlights what it is doing on this route map.

Most users of spatial data will either want to convert or transform between different coordinate reference systems, or they will want to carry out computations using the data that they have and wish to understand its geometry and limitations. If the reader is trying to understand more about the latter, then the point of reference is the section that introduces the coordinate type (mostly Chapter 2, but with projections covered in Chapter 3).

Chapters 2, 3, and 4 can therefore be thought of as the core of this book, providing the reference point for fundamental definitions of coordinate reference systems and operations. The remaining chapters and appendices aim to provide further context and supporting information. Chapter 5 is an introduction to global navigation satellite systems (GNSS) such as GPS, with special emphasis on the problems that are encountered when using these systems and trying to integrate the coordinate information that

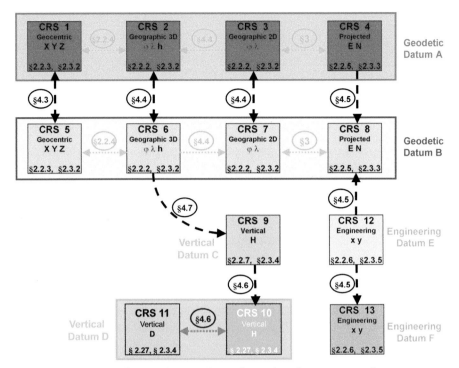

Figure 1.5 **Conversions and transformations between coordinate reference systems (CRSs).**

they provide into existing coordinate reference systems such as those used for national mapping. This is not an exhaustive coverage of the subject, but since positioning with systems such as GPS is a significant source of many of the problems that arise with coordinates and datums, this chapter aims to give enough of an overview of the principles of operation for readers to understand the consequences for the way in which GNSS-derived coordinates should be treated.

Chapter 6 is a series of case studies designed to reinforce the reader's understanding of the issues by exploring solutions to particular problems relating to coordinate reference systems and transformations.

Throughout the book we have maintained the use of the terminology recommended by the ISO 19111 standard. Readers may occasionally find that this clashes with some quite widespread colloquial usages in different sectors of the geomatics and mapping industries. To help overcome any confusion that may arise as a result, Appendix A provides a means of matching colloquial usages to the official ISO terms.

Finally, the last four appendices group together some of the formulae that are useful for computations within a coordinate reference system or for transforming from one coordinate reference system to another, where it has not been appropriate to incorporate these into the main text.

In general, however, the assumption has been made that most readers will have access to software that will carry out the actual computations for them. The emphasis is therefore less on the mathematics of computations and transformations, and more on being a practical guide to which operation should be used under what circumstances, and how to understand the result.

2

COORDINATES AND REFERENCE SYSTEMS

2.1 The Earth – geoid and ellipsoid

2.1.1 The geoid

With its mountains, valleys and plains, the Earth is an irregular body. Its surface is difficult to compute across and it is more convenient to use a shape close to sea level.

The shape of the Earth is given by the form of a surface that is everywhere perpendicular to the direction of gravity (such a figure is termed an *equipotential surface*). The force and direction of gravity are affected by irregularities in the density of the Earth's crust and mantle. It therefore follows that the form of an equipotential surface is somewhat irregular.

To a very good approximation, the form of the *mean sea level* surface is equipotential, since the sea would be expected to be perpendicular to gravity. In fact the seas and oceans contain permanent currents, which cause permanent slopes with respect to the direction of gravity.

The true shape of the Earth is known as the *geoid*, and this can now be defined as *that equipotential surface that most closely corresponds to mean sea level.* Worldwide, the difference between the geoid and mean sea level is at the most around 1 m (and changes slowly over wavelengths of tens of kilometres). For many purposes, the geoid and mean sea level can be considered synonymous. For some purposes, however, it is important to note that they differ.

In many situations it is necessary to have knowledge of the form of the geoid, either globally or for a specific region. This can be derived (with a great deal of effort) from gravity observations, observations of satellite orbits, satellite altimetry over the oceans, and other data sources. In developed countries with a dense network of gravity observations, the separation might be known to an accuracy of 5 or 10 cm or even better. In other parts of the world it would be much worse than this. The point of reference would be the national mapping organisation for the country concerned or a university research department.

The most common reason for needing to know the form of the geoid is when transforming between different height systems. A description of different models of the geoid that are available is given in section 4.7.

2.1.2 Models of the shape of the Earth – the ellipsoid

Because of the irregularities caused by local changes in density, the geoid remains a difficult surface to compute across. Computations are considerably simplified by making them on a model of the Earth. The geoid is approximately spherical, with a radius of about 6370 km. For many low accuracy or small scale[1] mapping applications, a sphere is a satisfactory model of the Earth.

A rather better approximation of the shape of the Earth is an *ellipsoid of revolution*, often more conveniently called *an ellipsoid*. This surface is formed by an ellipse that has been rotated about its shortest (the minor) axis, or by 'squashing' a sphere at the poles. This *flattening* of the poles is slight, about 22 km in 6378 km. An ellipsoid with this degree of flattening would, to the eye, appear indistinguishable from a sphere and it must therefore be emphasised that for purposes of clarity most diagrams of the ellipsoid exaggerate the flattening. The term *spheroid* can also be used for this shape, and although some would differentiate between an ellipsoid and a spheroid (see Appendix A), in this context the two can be regarded as synonymous. Despite the flattening being slight, computations on the ellipsoid give significantly different results compared to those on the sphere – this is discussed in Chapter 3 – and, except for low accuracy applications, an ellipsoid should always be used as the model of the Earth.

An ellipsoid formed in this way is referred to as an *oblate ellipsoid*, as opposed to a *prolate ellipsoid*, which is formed by rotating about the major axis thereby extending the distance between the poles compared to the diameter at the equator. As all the ellipsoids referred to in this text will be oblate, the term ellipsoid will be used without ambiguity.

With reference to Figure 2.1, an ellipsoid is defined by the size of two parameters:

the *semi-major axis*, a

the *semi-minor axis*[2], b

From these two parameters, it is possible to derive further parameters. Thus:

Flattening, f, is defined as $f = \dfrac{a - b}{a}$ (2.1)

but often expressed as *inverse flattening*, 1/f.

Eccentricity, e, is defined as $e^2 = \dfrac{a^2 - b^2}{a^2}$ (2.2)

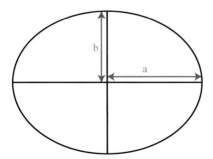

Figure 2.1 Defining parameters of an ellipsoid.

---

[1] Scale is a ratio. Small scales have a large denominator such as 1 000 000 and maps at small scales cover large areas. Map sheets at large scales, typically between 1:500 and 1:10 000, cover small areas – a few square kilometres.

[2] For a triaxial ellipsoid, the conventional symbol for the semi-median axis is b, with c for the semi-minor axis. As the Earth can be modelled as an ellipsoid of revolution in which the semi-median axis equals the semi-major axis, the symbol b is applied to the semi-minor axis.

Furthermore, e, f, and b can be related to each other as:

$$e^2 = 2f - f^2 \tag{2.3}$$

$$\sqrt{1 - e^2} = (1 - f) = \frac{b}{a} \tag{2.4}$$

Thus, an ellipsoid can be completely defined using the two parameters (a and b) or (a and f) or (a and e) and the remaining parameters can be found as necessary. Although historically models of the Earth were described using a and b, it is now more usual to use a and 1/f. A difficulty with using 1/f rather than f is that for a sphere 1/f is undefined, since the flattening is zero.

Many hundred estimations of the best-fitting ellipsoidal model of the Earth have been made over the past two centuries. Of these most have been derived for scientific purposes, but a few tens of these models remain in daily use for mapping.

The best estimate has been significantly refined with the advent of Earth-orbiting satellites. The accepted value today for the best-fitting model was adopted by the International Union of Geodesy and Geophysics at its meeting in Canberra in 1979 as part of the Geodetic Reference System of 1980, normally abbreviated to GRS 1980 (Moritz 1988). Parameters for some frequently used ellipsoids are tabulated below.

A tabulation of parameter values for other ellipsoids used in modern surveying and mapping is included in the EPSG Geodetic Parameter Dataset (OGP 2007a).

Due to a slight difference in the gravity terms adopted in the definition of the GRS 1980 and WGS 84 systems, there is a difference in their ellipsoid flattening values that amounts to (Hofmann-Wellenhof *et al.* 2001):

$$\Delta f = f_{GRS80} - f_{WGS84} = 16 \times 10^{-12} \tag{2.5}$$

For most practical purposes, this difference is insignificant and these two ellipsoids can be considered equivalent. However, the differences in both size and shape between these two ellipsoids and other older figures are significant and cannot be ignored.

How good a fit to the actual shape of the Earth are these ellipsoidal models? Figure 2.2 is a plot of the height of the geoid above the WGS 84 ellipsoid, and it can be seen that

Table 2.1 Ellipsoid parameters.

Ellipsoid name	Semi-major axis (a)	Flattening (f)	Comment
GRS 1980	6 378 137 m	1 / 298.257222101	The international standard.
International 1924	6 378 388 m	1 / 297.0	A former international standard.
GRS 1980 Authalic Sphere	6 371 007 m	0	An authalic sphere is one with a surface area equal to the surface area of the ellipsoid.
WGS 84	6 378 137 m	1 / 298.257223563	Used by the GPS satellite navigation system.

Figure 2.2 The geoid with respect to the WGS 84 ellipsoid. The heights have been obtained from the EGM96 model of the geoid (NGA 2007).

the differences between the two are mostly contained within the band ± 100 m, with a root mean square (rms) value of around 30 m. The largest differences generally have quite a long wavelength, and so we see features such as a high of around +65 m covering much of the North Atlantic, a low of around –90 m covering much of the Indian Ocean, and so on. A positive difference means that the geoid is above the ellipsoid; with a negative difference the geoid is below the surface of the ellipsoid. However, note that these differences of up to 100 m are tiny when compared with the 22 km difference between the ellipsoid and the sphere. The ellipsoid is *visually* indistinguishable from the sphere; the difference between the geoid and the ellipsoid is even less easy to depict, and once again all conventional diagrams that show the geoid waving up and down with respect to the ellipsoid are grossly exaggerated for clarity.

2.2   Coordinate systems

2.2.1 Coordinate system attributes

Every coordinate system has a set of attributes that enables users to interpret the meanings of the numerical values of the coordinates. Attributes of any coordinate system are:

- The dimension of the coordinate system. The dimension defines the number of axes associated with the system. At every point, one coordinate is associated with each axis, so the dimension also defines the number of coordinates describing the position of a point.

Then for each axis of a coordinate system it is necessary to define certain attributes:

- Name and/or abbreviation for the axis.
- The sequence of the axes. Coordinates are listed in this order.
- The direction in which coordinates increment along the axis.
- The units for measurements on the axis.

If any of the attributes are changed, then the coordinate system is changed.

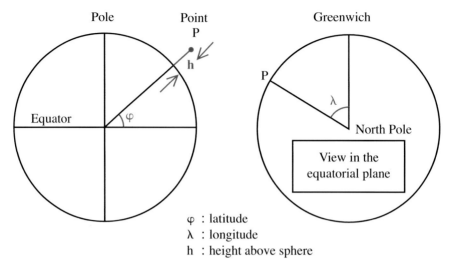

φ : latitude
λ : longitude
h : height above sphere

Figure 2.3 **Coordinate system for a sphere.**

2.2.2 Coordinate systems for the sphere and ellipsoid

Three-dimensional coordinates can be defined with respect to a sphere as:

Latitude: the angle north or south from the equatorial plane.
Longitude: the angle east or west from an identified meridian.
Height: a distance above the surface of the sphere.

The definition of latitude is a natural one, in that the poles are clearly defined points on the surface of the sphere, and the equator is the circle that bisects the two. On the other hand, the definition of longitude is to some extent arbitrary, as there are no physical reasons for choosing a particular meridian as the reference. That chosen is called the *prime meridian*. Historically, different prime meridians were used by different countries, but the convention of using Greenwich is now the international standard. The identification of the chosen prime meridian is part of a datum definition, discussed in section 2.3.

In conjunction with this coordinate system, it is useful to define the following terms:

Parallels: lines on the surface of the sphere parallel to the equator. These are lines of equal latitude.
Meridians: lines on the surface of the sphere running from pole to pole. These are lines of equal longitude.

Parallels and meridians intersect each other at 90°. A lattice of parallels and meridians is known as a *graticule*. Figure 2.4 shows the pattern that results from parallels and meridians depicted at intervals of 5°.

A similar system for depicting position can be employed on the ellipsoid. The coordinates are defined as *latitude, longitude*, and *ellipsoidal height*. These are shown in Figure 2.5. The adjective ellipsoidal is applied to height to distinguish it from height above the geoid, discussed in section 2.3.4. Referring back to Figure 2.3, it will be recalled that latitude is an angle measured at the centre of the sphere. Putting this another way, the latitude of a point on the surface of the sphere is determined by

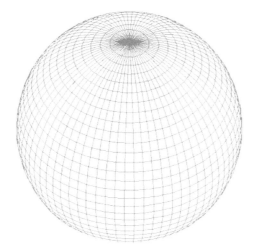

Figure 2.4 **A graticule of parallels and meridians.**

the intersection with the equatorial plane of a line from the point in question that is perpendicular to the surface of the sphere at that point. This same principle applies to an ellipsoid, but in the ellipsoidal case the intersection will generally not be at the exact centre of the ellipsoid. It can be seen by inspection of Figure 2.5 that the latitude and longitude are defined with respect to the direction of the *ellipsoidal normal*, a line from the point in question that is perpendicular to the surface of the ellipsoid at that point.

Coordinates defined in this way are known as *ellipsoidal coordinates*, and are the basis for all mapping systems. An ellipsoidal coordinate system may be two-dimensional (latitude and longitude) or three-dimensional (latitude, longitude and ellipsoidal height). Note that, like latitude or longitude, ellipsoidal height cannot exist on its own but only as part of an ellipsoidal coordinate system.

The term ellipsoidal coordinate system applies regardless of whether the model is an ellipsoid or a sphere. In this context, the sphere is taken to be a degenerate

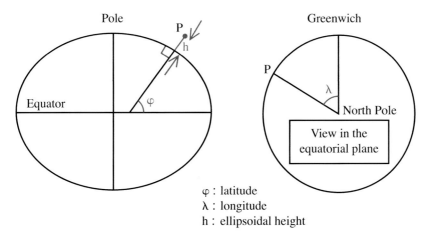

φ : latitude
λ : longitude
h : ellipsoidal height

Figure 2.5 **Ellipsoidal coordinate system.**

ellipsoid. A *spherical coordinate system* is something different: it is a three-dimensional coordinate system with one linear and two angular axes from a common origin at the centre of the model, sometimes used in space applications. In an ellipsoidal coordinate system, the axes do not have a single origin: the height is measured from the surface of the figure.

Attributes of ellipsoidal coordinate systems

- Dimension: an ellipsoidal coordinate system may be two-dimensional (latitude and longitude) or three-dimensional (latitude, longitude and ellipsoidal height).
- Abbreviations for the axes. The symbols φ, λ and h are conventionally used for latitude, longitude and ellipsoidal height, respectively.
- The sequence of the axes. For an ellipsoidal coordinate system the conventional order of coordinates is latitude, longitude and (in the three-dimensional case) ellipsoidal height.
- Axis direction: for latitude, angles measured northward from the equator conventionally are treated as positive; angles southward from the equator are negative. For longitude, angles measured eastward from the prime meridian are positive; angles westward from the prime meridian are negative. For height, distances measured up from the surface of the ellipsoid are positive; distances down from the surface of the ellipsoid are negative.
- Axis units: latitude and longitude are typically expressed in degrees (1/360th of a circle). For computer applications, the degrees are decimal numbers, but for data presented to the human reader sexagesimal degrees – degrees, minutes and seconds with hemisphere indicated, for example 53°23'36.5"N, is more appropriate. In France, grads (1/400th of a circle) may be used. For some surveying applications, longitude may be expressed in hours (1/24th of a circle). It can be seen that 1 hour is equivalent to 15 degrees. Ellipsoidal height is typically expressed in metres.

It is possible to carry out computations in an ellipsoidal coordinate system, particularly in the region close to the ellipsoidal surface. In traditional land surveying, these would have been used when computing coordinates in extensive geodetic surveys, a process that has largely been superseded by the use of geocentric Cartesian coordinates (section 2.2.3) and satellite navigation systems (Chapter 5). Over smaller areas it has usually been possible to use a map grid. One of the few remaining applications where computations are needed in ellipsoidal coordinates is in the definition of boundaries between states, or between mineral concessions. For this reason, a summary of some of the most useful formulae and procedures is given in Appendix C. Otherwise, any textbook on geodesy, such as Bomford (1980) or Torge (1991), will have further and more detailed information on ellipsoidal geometry.

For many applications that do not require the highest accuracy, the sphere is an adequate representation of the Earth. Appendix B summarises some of the formulae that can be used in conjunction with the spherical model, such as finding the distance and azimuth between points of known latitude and longitude.

2.2.3 Geocentric Cartesian coordinates

The formulae involved in computations based on ellipsoidal coordinates are complex, and entirely inappropriate when considering observations made to or from satellites.

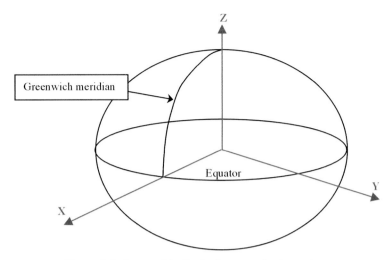

Figure 2.6 **Geocentric Cartesian coordinate system.**

More appropriately, a set of Cartesian axes is defined with its origin at the centre of the ellipsoid. The Z axis is aligned with the minor axis of the ellipsoid (the 'polar' axis); the X axis is in the equatorial plane and aligned with the prime meridian; the Y axis forms a right handed system (see Figure 2.6). Such a system is often referred to as *geocentric*, due to its origin being placed at the centre of the Earth. This nomenclature is not strictly correct: the coordinate system origin is at the centre of the *model of the earth*. As we shall see when we discuss datums, the centre of the model may not coincide with the centre of the real Earth.

Many of the attributes for geocentric Cartesian coordinate systems are implicit in the system definition. The system is by definition three-dimensional. Conventionally, the axes are ordered X, Y and Z. For geodetic applications, the axes' units are metres; some space applications may use kilometres.

2.2.4 Conversion between ellipsoidal and geocentric Cartesian coordinates

Geocentric Cartesian coordinates may be converted from and to ellipsoidal coordinates by a set of formulae that require knowledge of the parameters of the ellipsoid (section 2.1). The semi-major axis for the ellipsoid (a) must be in the same length units as the Cartesian axes.

$$X = (v + h) \cos\varphi \, \cos\lambda \qquad (2.6)$$

$$Y = (v + h) \cos\varphi \, \sin\lambda \qquad (2.7)$$

$$Z = \{(1 - e^2) \, v + h\} \sin\varphi \qquad (2.8)$$

where   φ is the latitude, positive north;

λ is the longitude, positive east;

h is the ellipsoidal height (the height above the ellipsoid surface).

ν is the radius of curvature in the prime vertical, described in Appendix C.2

The reverse computation is also possible, in which ellipsoidal coordinates are found from geocentric Cartesian by:

$$\tan \lambda = \frac{Y}{X} \tag{2.9}$$

$$\tan \varphi = \frac{Z + \varepsilon\, b\, \sin^3 u}{p - e^2 a\, \cos^3 u} \tag{2.10}$$

$$h = p \sec \varphi - \nu \tag{2.11}$$

where a and b are the semi-major and semi-minor axes of the ellipsoid (see equation 2.1) and

$$p = (X^2 + Y^2)^{\frac{1}{2}} \tag{2.12}$$

$$\tan u = \frac{Z}{p}\frac{a}{b} \tag{2.13}$$

$$\varepsilon = \frac{e^2}{1 - e^2} \tag{2.14}$$

and all other terms have been defined above.

2.2.5 Map projection coordinates

Before the advent of digital computing, computations in two-dimensional ellipsoidal coordinates (latitude and longitude) were non-trivial. Far easier were computations in a two-dimensional Cartesian system. A Cartesian coordinate system has axes that are straight and mutually perpendicular. Various mathematical techniques were evolved for mapping ellipsoidal coordinates into Cartesian coordinates. This is the science of map projections, discussed in Chapter 3.

Map projection coordinate system attributes

There are few globally-adopted conventions for the attributes of Cartesian coordinate systems used with map projections.

- Axis names: usually reflect the direction in which coordinates increment, normally easting and northing but westing and southing are sometimes encountered in older systems.
- Axis abbreviations: usually the first letter of the axis name, E or N (or S or W). These may not necessarily be in English; for example in Portuguese they may be encountered as P (south) and M (west). However, X and Y or x and y are frequently encountered, but there is no single convention as to which direction these are associated with. In any particular system, X may be defined to be the easting but it could equally well be defined to be the northing, with Y as the other axis.
- Axis order: when X and Y are used as abbreviations, the axis order is almost invariably X, Y. Whether that means easting, northing or northing, easting

depends upon the definition for the system. When the axis abbreviations reflect the positive direction of the axes, again there is no global convention: some systems use E, N whilst others use N, E.

- Axis units: usually metres. But in the United States, there is frequent use of the foot. Units are a science in their own right (metrology). There are different types of metres and feet, which differ in length by small, but in mapping applications significant, amounts. The metre is now defined (in the SI system) with respect to a physical phenomenon – the length travelled in a vacuum by light during 1/299792458th of a second (BIPM 1983). This is known as the ISO[3] or *international metre*, abbreviated as *m*. ISO defines a foot as 0.3048 exactly international metres – this is the *international foot* (*ft*). But historically the length was that of a standard bar. Due to imperfections in manufacture and the effects of changes over time, 'standard' bars were not consistent with each other and different bars were legally adopted in different jurisdictions. So there was a French Legal Metre and a Prussian Legal Metre. The latter, in the form of the German Legal Metre, is still found in use in some coordinate systems in Germany and German former colonies such as Namibia; 1 GLM = 1.0000135965 m. The situation with imperial measure is even more confusing, as some standard foot bars have been compared to standard metre bars on a number of occasions, with different determinations of the conversion ratio. Various mapping systems have then adopted a particular conversion ratio. In the United States, a foot has been legally defined as 12/39.37 metre exactly (1 ftUS = 0.3048006096… m). This unit – known as a *US Survey foot* – has been adopted in some States, but other States have adopted the international foot. These small differences in unit length become significant in map projections because coordinates are usually calculated from the equator or from the apex of a cone well outside of the Earth, far away from the area being mapped. When the distance is several million units, small differences become significant.

One further point associated with map projection units is that for computations the linear parameters for the ellipsoid and the map's coordinate system must all be in the same length units.

2.2.6 Cartesian coordinates for engineering applications

For some applications, the area being measured is sufficiently small for the curvature of the Earth to be minimal and therefore for projections to be considered an unnecessary inconvenience. In such situations a Cartesian (XYZ) system may be used, but one that is rather different in nature to the geocentric ones introduced in section 2.2.3. The X and Y axes will usually be oriented in the horizontal plane and, if a normal survey convention is used, the Y axis will be a proxy for the direction of 'north', from which angular measurements will be expressed. However, this will usually be an approximate direction or in other cases entirely unrelated to the actual direction. The Z axis is defined as the 'up' direction of the local vertical.

In most respects, therefore, this Cartesian system is – as a system – just like the geocentric ones already encountered. It just happens to be located at a very different point, usually on the surface of the Earth, and orientated in a completely different way. There is a subtle difference, however, in that the Z direction is in theory not the

---

[3] ISO, the International Organization for Standardization.

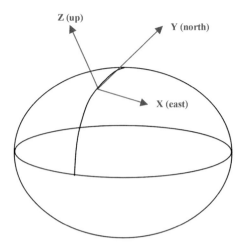

Figure 2.7 Topocentric coordinate system.

same at all points, as it varies with the curvature of the Earth. However, such a system would usually be employed on construction sites less than a kilometre or so in extent, and the assumption of a flat Earth would be acceptable at the accuracies required (although an exception would be a project to construct a large particle accelerator, for example, where the accuracy requirements are more exacting). These coordinate systems are sometimes called *topocentric*.

A rather different type of engineering application would involve measurements on a man-made moveable object such as a ship or an aircraft. To position an acoustic transducer with respect to an inertial sensor and a GPS antenna, for example, might involve a small survey being carried out on a vessel. In these situations, Earth curvature is of no relevance at all, and the most appropriate coordinate system is again a three-dimensional Cartesian one. Usually the axes would be aligned with respect to some natural features of the object being measured, such as longitudinal axis, wingspan, and so on.

The attributes of all these systems may be stated in exactly the same way as for the geocentric Cartesian systems. They are two- or three-dimensional, with axes labelled (usually) xy or xyz. The units are generally metres. Where they differ is in how they are referenced to the body on which they are used, and this is covered in section 2.3.5.

2.2.7 Gravity-related systems (height and depth)

The complexities of the geoid surface make it unsuitable for computation across the surface. But the geoid is a suitable surface to which height or depth measurements can be related. The geoid is an equipotential surface perpendicular to gravity. It is therefore natural for measurements from the geoid to be along the direction of gravity. A one-dimensional coordinate system with measurements along the direction of the gravitational field is called a *vertical coordinate system*. When the coordinates increment upwards (towards lower gravitational attraction), the coordinates are *gravity-related heights*, usually given the symbol H. If the coordinates increment in the direction of gravitational attraction, i.e. downwards, the coordinates are *gravity-related depths*, more conveniently referred to as *depths*. As with all coordinate systems, the units for the vertical coordinate system need to be identified.

Gravity-related height is a non-scientific term embracing a number of scientific concepts including orthometric height, normal height and geopotential number. These similar concepts differ from each other in the assumption made about the strength of the gravity field and the shape of the geoid. It should be emphasised, however, that for most users the subtleties in definition of these types of gravity-related height are not relevant.

The use of corrector surfaces to transform between ellipsoidal heights and gravity-related heights is covered in section 4.7.

2.2.8 Miscellaneous coordinate systems

Two-dimensional Cartesian coordinates are often used in a local system. But it may be more convenient to use a polar coordinate system. A useful example of this is the two-dimensional coordinate system shown in Figure 2.8.

In this example, the coordinates of the point P may be quoted in either the rectangular form (X,Y) or the polar form (r, θ): changing from one to the other is a relatively straightforward procedure that does *not* involve changing the position of the origin, nor the orientation of the system. Note that in Figure 2.8 the polar angular coordinate is measured as a clockwise bearing from north. This is conventional in surveying applications and differs from the mathematical convention of measuring anticlockwise from the abscissa.

Ellipsoidal, vertical and Cartesian coordinate systems are the most frequently encountered in GIS, remote sensing, and surveying. But in addition to these and the polar system described above, other types of coordinate system may be required in special circumstances:

- *Affine*: straight axes but, unlike a Cartesian system, the axes are not necessarily orthogonal. Raw remote sensing data may be acquired in such as system.
- *Cylindrical*: a polar system extended into 3 dimensions by the addition of a straight axis perpendicular to the plane of the polar system.
- *Spherical*: a 3-dimensional coordinate system with one linear and two angular measurements from a common origin. Sometimes used in space applications, when the origin is geocentric. Not to be confused with an ellipsoidal coordinate system applied to a sphere: in an ellipsoidal coordinate system the axes do not have a single origin: the height is measured from the surface of the figure.

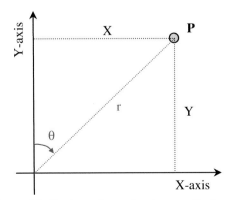

Figure 2.8 Rectangular and polar coordinates.

- *Linear*: distances are measured along a curvilinear object. Positions of other features near to the curvilinear object may sometimes be described through perpendicular offsets from the curvilinear object. This is sometimes used in the construction of railways or pipelines, for example, where the coordinate is referred to as 'chainage' or 'kilometrage'.

2.3   Datums and coordinate reference systems

2.3.1 Datum overview and classification

A datum is the information that is required to fix a coordinate system to an object. In most cases in the context of this book, the object concerned is the Earth. The information required will depend on the nature of the coordinate system being used and the number of dimensions that it encompasses, but in general terms a datum positions and orientates a coordinate system to turn it into a *coordinate reference system*.

Several types of datum may be identified:

- *Geodetic* datum – a datum describing the relationship of coordinate systems for an ellipsoidal (or spherical) model of the Earth with the real Earth.
- *Vertical* datum – a datum describing the relationship of gravity-related heights with the Earth.
- *Engineering* datum – a datum that describes the relationship of a coordinate system to a local reference. The reference can be any moving object, such as a vehicle, an aircraft, or a ship as illustrated in Figure 1.2. But it can also be a small site on the Earth, for example a construction site or an industrial plant. In general, we can say that an engineering datum is any that is not specifically geodetic or vertical.
- *Image* datum – a type of engineering datum that describes the relationship of a coordinate system to an image. It is distinguished from an engineering datum through having attributes describing the origin of a cell-based coordinate system.

Each of these is described in the following sections.

2.3.2 Geodetic datums and coordinate reference systems

2.3.2.1 Datums

Let us first look at the situation with regards to an ellipsoidal coordinate system to be related to the Earth. As we saw in section 2.1, the shape of the Earth is represented by the geoid and the preferred model for the shape of the Earth is an ellipsoid, with a graticule of latitude and longitude coordinates forming an ellipsoidal coordinate system. To relate the ellipsoidal coordinate system to the Earth, we need to relate the ellipsoid to the geoid. A *geodetic datum* is the mechanism through which this relationship between ellipsoid and geoid is defined.

Before the advent of earth-orbiting satellites, the classical technique for defining a geodetic datum was to *choose* an ellipsoid – either one that was known to fit well with the local geoid, or that was recognised at the time as being the best representation of the size and shape of the Earth – and to define the position and orientation of this ellipsoid at a chosen place. This point is the datum origin and is sometimes called the fundamental point. At this point, the geoid-ellipsoid separation, N, and the alignment of the ellipsoidal normal are chosen, usually as zero for the separation and parallel to gravity for the normal. The alignment of the ellipsoid normal is defined through the *deviation of the vertical* (an explanation of this term is given in section 2.3.4.2). This has the effect of fixing the chosen ellipsoid to the geoid (usually coincident and parallel) at the point of origin. The orientation of the minor axis of the ellipsoid is made parallel to the rotation axis of the Earth by making *Laplace azimuth observations*: observations of astronomical azimuth (with respect to the stars and the pole of the Earth's rotation) are converted to geodetic azimuth (defined on the ellipsoid) by a formula that forces the poles of the two systems to be the same. An example of this classical technique for geodetic datum definition is the European datum of 1950 (ED50).

The meridian for zero longitude – the *prime meridian* – also has to be chosen and is part of the datum definition. After the advent of wireless telegraphy in the late 19th Century, the meridian through the old astronomic observatory at Greenwich near London in the UK was agreed as the international standard. Prior to this it had been customary to choose the meridian that passed through the local observatory. Local observatories may still be used as the fundamental point, but nowadays they will be assigned a non-zero longitude value such that longitudes are relative to the Greenwich meridian. An example is the observatory at Uccle, used as the origin for the Belgian datum of 1972 and assigned a longitude value of 4°21'24.983" East of Greenwich. Although increasingly rare, it is still possible to encounter datums using a prime meridian where the national observatory is assigned zero longitude. For example, until the very recent adoption of its new datum, France had continued to use the meridian through Paris as its prime meridian.

The situation is shown diagrammatically in Figure 2.9. The ellipsoid in black represents the International 1924 ellipsoid adopted for European Datum 1950, or ED50. Note that although this ellipsoid fits well with the geoid in northern Europe, when it is so fitted it then does not fit at all well in some other parts of the world.

However with a different origin position and alignment, the same International 1924 ellipsoid may be fitted in another part of the world. For example, for the 1956

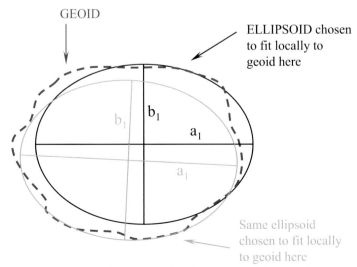

Figure 2.9 **An ellipsoid fitted locally to the geoid.**

Provisional Datum of South America (PSAD56), the International 1924 ellipsoid was defined to coincide with the geoid at La Canoa in Venezuela. The same principles for defining orientation and longitude were adopted as for ED50. When used for PSAD56, the ellipsoid no longer fits to Europe. This is illustrated by the green ellipsoid in Figure 2.9.

Of course, it is not necessary to adopt any ellipsoid in particular and not all datums have been defined to use the International 1924 figure. In Great Britain the ellipsoid adopted was the Airy 1830, developed by a former Astronomer Royal, whilst for its 1927 datum the US adopted the Clarke 1866 ellipsoid. For the German national network, the Bessel ellipsoid of 1841 was adopted; the datum origin was at Potsdam, later to be used as the origin for ED50. Prior to the advent of satellite geodesy, all national datums were of necessity defined independently of each other.

The effect of defining the datum in this way is that the ellipsoid is not exactly geocentric; its centre is offset from the centre of the Earth. Different datums will have differing offsets.

Since the advent of earth-orbiting satellites, the technique for relating the ellipsoid to the Earth has changed. Firstly, the size and shape of the best-fitting ellipsoid is now much more refined, with the GRS 1980 figure accepted as the best model. Secondly, a system known as the International Terrestrial Reference System (ITRS) is adopted. This concept describes a three-dimensional geocentric Cartesian coordinate system (see section 2.2.3 and Figure 2.6) in which the direction of the Earth's rotation axis and its speed of rotation are determined by the International Earth Rotation and Reference Systems Service, IERS (McCarthy and Petit 2003). The IERS was established in 1998 to replace the International Polar Motion Service and the Earth rotation functions of the Bureau International de l'Heure. The wider remit of the IERS is to establish both the terrestrial reference frame and the international celestial reference system (a coordinate system that does not rotate with the Earth), and to determine the connections between the two systems. This process involves monitoring and predicting the rotation of the Earth and the movement of its poles, and is essential information if, for example, a determination of a satellite's position is to be given in a terrestrial system.

In these global datums, the prime meridian adopted is the international reference meridian defined by the IERS. This definition includes the definition of Universal Time and allows for movement of the polar axis of the earth – precession and nutation – and although in principle coincident with the meridian through the old Greenwich observatory in practice does not exactly agree (Ordnance Survey 1999). The variation of approximately 102 m may be significant for many studies. For practical purposes in the fields of GIS and remote sensing, the use of the term 'Greenwich prime meridian' implies the international reference meridian defined by the IERS.

The origin of the ITRS is by implication at the mass centre of gravity of the Earth (the centre of the geoid). The ITRS does not need an ellipsoid but when its Cartesian coordinates are converted to ellipsoidal coordinates the GRS 1980 ellipsoid is used. The situation is illustrated in Figure 2.10.

Putting Figures 2.9 and 2.10 together as Figure 2.11, we can see that multiple models of the Earth (ellipsoids) exist, and some of these models are related to the geoid through more than one datum definition. Almost by definition, an ellipsoid fitted to the geoid locally approximates the geoid in the region much more closely than does an ellipsoid fitted regionally or globally.

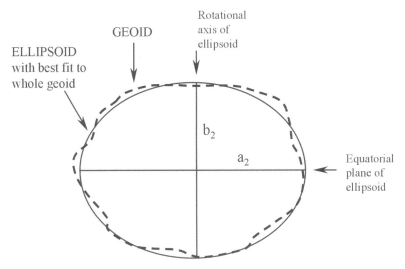

Figure 2.10 **An ellipsoid fitted globally to the geoid.**

It is worth noting from Figure 2.11 that the orientations of the chosen ellipsoids may differ and that the centres of the various datums are in general not coincident. Geocentric transformations are discussed in section 4.3.

Note also that the geodetic datum *definition* does not depend upon its offset from another geodetic datum. We will discuss this further in section 4.6.3 when we describe the transformation hub concept.

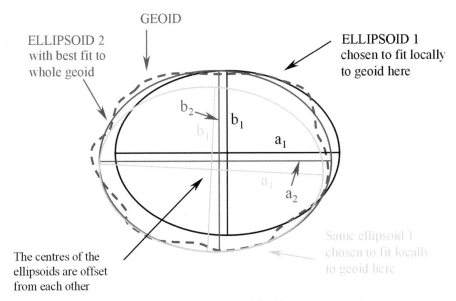

Figure 2.11 **The Earth (the geoid) with several models.**

2.3.2.2 Coordinate reference systems

We have seen that the position and orientation of an ellipsoidal or geocentric Cartesian coordinate system is related to the Earth – more correctly, related to a model of the Earth – through a geodetic datum. The origin of the geodetic datum is defined. But before it can be used it is necessary to *realise* the datum; that is, to provide physical monuments of known coordinates throughout the area of interest. Any mapping or surveying that is carried out is then related to the coordinates of those control points that are available locally, rather than to the rather abstract concept of a datum defined at an origin point.

Before the age of electronic distance measuring devices and satellite receivers, the surveys carried out to realise the coordinates of the physical monuments were often colossal undertakings. In India, for example, starting in 1800 the entire sub-continent had a geodetic triangulation framework established over the course of the 19th Century (Keay 2000). An initial baseline of around 10 km in length was measured using calibrated rods – any error would be propagated proportionally across the rest of the survey, so this had to be a very accurate measurement. Building from the ends of the measured baseline, the coordinates were carried across the country by measuring the angles in a chain of triangles, with occasional check measurements being made with additional baselines (each one taking several months of painstaking observations). All observations had to be adjusted (without the aid of an electronic computer) to determine the coordinates of the control monuments. The survey of India was an extreme case due to its size, but was in most ways typical of the means used to establish geodetic control for mapping. Of course, the observations made in such cases were not perfect and the errors affected the coordinates determined for the control points when compared to their 'true' position in the reference system. However, having published the coordinates, these then became the basis for subsequent mapping and hence *the coordinates, although in error, were now an essential part of the reference system.*

Were there to be a later series of survey observations, these could be based on the same datum origin but the new adjustment[4] of the observations would give rise to a second set of coordinates of the monuments. This would be a second realisation of the datum. In this case, both initial and subsequent realisations share the same coordinate system, ellipsoid, prime meridian, and datum origin, but away from the origin the coordinates of positions are different. When there are subsequent realisations, they are often differentiated from the initial realisation by the inclusion of the year in which the adjustment was made. An example would be the triangulation of Great Britain referred to as OSGB 1936 or OSGB36 being re-computed with additional distance and satellite observations and referred to as OS(SN)80 (Ashkenazi *et al.* 1986). The 'SN' refers to this being a scientific network that was not actually adopted as the official working set of coordinates.

The datum name may describe the area covered by the realisation, as in the European Datum of 1950 and the South American Datum of 1969. Alternatively the datum realisation may be named after the origin point rather than the area covered by the realisation, as in the Bogotá datum used historically throughout Colombia.

A *coordinate reference system* (CRS) then is a coordinate system that has been referenced to an object through a datum. Here, *datum* is the full datum realisation.

---

[4] The term *adjustment* refers to computing by mathematical and statistical principles the most probable coordinates from the original observations. It is in common use in surveying and geodesy, but to the uninitiated sounds dangerously close to *fudging*.

The coordinate reference system name is usually that of the datum realisation. The use of datum name as a coordinate reference system name tends to reduce the visibility of the coordinate system component. It is however critical that the coordinate system attributes are known, as these describe the order and units of the coordinates.

Coordinate reference systems can be classified as being of various subtypes, depending upon the type of datum. A coordinate reference system with a geodetic datum is a *geodetic coordinate reference system.*

The distinction between datum realisation and coordinate reference system is rather subtle, and largely a data modelling convenience. A datum cannot be realised without coordinates. But for many geodetic purposes when discussing datum realisation, it is adequate for the coordinates to exist, with detailed knowledge of the coordinate system attributes not being necessary. There is some ambiguity in the use of the word datum. In some contexts it refers explicitly to the arrangements at the origin. More usually it refers to both the origin definition and the realisation. It is preferable to avoid this ambiguity by using the term coordinate reference system.

The requirement to realise the datum through measurements that are imperfect then presents the problem where the coordinate reference system includes all the measurement errors and computational approximations of the control survey. These surveying errors inherent in coordinate reference systems cause complications in transforming between systems. This is discussed further in chapter four.

The distinction between *coordinate reference system* and *coordinate system* – the inclusion of datum – is often not well understood. Sometimes in inexact colloquial usage the term *coordinate system* is given when *coordinate reference system* is meant. Appendix A gives cross references between frequently used colloquial terminology and the formal nomenclature of the ISO used here.

The terms *geographic coordinate reference system* and *geocentric coordinate reference system* are in common usage in the surveying, GIS, and remote sensing communities. In these communities, a geographic CRS is a geodetic coordinate reference system in which the coordinate system is ellipsoidal, whilst a geocentric CRS is a geodetic coordinate reference system in which the coordinate system is geocentric Cartesian. Neither of these terms is completely satisfactory. Geographic coordinates will usually be geodetic, that is related to a geodetic datum, but in some circumstances latitude and longitude can be astronomic coordinates; that is, coordinates determined through astronomic rather than geodetic observations. Similarly, geocentric coordinates will usually be in a Cartesian coordinate system, but for space applications a geocentric spherical coordinate system is used.

2.3.2.3 Realisation of the ITRS – the International Terrestrial Reference Frame

The International Terrestrial Reference System (ITRS) is realised through a coordinate set published by the International Earth Rotation Service (IERS). The coordinate set is known as the International Terrestrial Reference Frame (ITRF). The concept and definition is the reference system, the realisation is the reference frame. But due to the Earth's plate tectonic motion, the relationship between the system and the frame changes slowly with time. IERS has published several reference frames, identified by the year in which the coordinates applied, for example ITRF88 (the first realisation), ITRF89, ITRF2000, and the latest realisation, ITRF2005. Plate tectonic motion means that the station coordinates are valid only at a defined epoch (first of January of the year of realisation), and the reference frame consists not only of the coordinates but also the velocities of the stations. It is then possible to compute station coordinates at some other epoch. The epochs are described as a fraction of the year: 1st January 2002 is 2002.0 and 1st June 2005 is 2005.42. It is then possible to refer to a reference frame at a chosen epoch, not necessarily in the year of realisation, for example ITRF2000 at epoch 2005.42.

The stations of the ITRF are rather thinly spread around the world, and for practical purposes it is necessary to densify this network on a regional or national basis. For example, the Australian densification, known as the Geodetic Datum of Australia 1994 or GDA94, is based on ITRF92 at epoch 1994.0.

On a local basis, it is generally inconvenient to have the coordinates of the reference frame continually changing with time. Instead the coordinate reference system may be related to the ITRF at a particular epoch but then fixed relative to the tectonic plate upon which the station set is placed; that is, the station velocities are zero by definition. This is the case in Europe, where ETRS89 is defined as having the same coordinates as ITRF89 at epoch 1989.0 but, unlike ITRF, moves with the European plate. ETRS89 is realised through a reference frame, ETRFxx, such as ETRF89. Since 1989, ETRS89 has diverged from the ITRF by approximately 2.5 mm per year. ETRS89 may then be further densified on a national or sub-regional basis. An example is IRENET95 covering the island of Ireland.

Note that there is a particular problem with nomenclature for the ETRS system and associated frames. ETRS89 is sometimes referred to as EUREF89 after the committee that defined it, whilst ETRS89 and ETRFxx are often treated as synonyms. Commonly the system and its realisation are both called ETRS89; we shall follow this practice throughout this book.

2.3.2.4 WGS 84

A geodetic coordinate reference system of especial interest is the World Geodetic System of 1984, or WGS 84. This is because it is the reference for positions determined through the use of the GPS navigation system.

The starting point for positioning using GPS is the satellite ephemerides (or orbital positions) that are broadcast by the system. These in turn have been determined from a set of monitoring stations maintained by US military authorities. It is the coordinates of these monitoring stations that effectively realise the WGS 84 coordinate reference system. So whilst in principle the conceptual definition of WGS 84 is, like ITRS, geocentric, its realisation differs from ITRF.

WGS 84 was previously only defined at an accuracy level of around 70 cm with respect to the ITRF. It was re-defined in 1994, however, in a form that was compatible with ITRF92 within 10 cm (Hooijberg 1997). Strictly speaking this realisation was then referred to as WGS 84 (G730), with 730 being the GPS week number in which the change was effected (the first full week of January 1994), but in practice G730 is often omitted. WGS 84 positions from before and after the January 1994 realisation will therefore differ by almost 1 m.

Since 1997, WGS 84 has been maintained to be consistent to a few centimetres with the then current ITRF. To achieve this there have been other re-definitions since that of 1994, such as the WGS 84 (G873) and WGS 84 (G1150) realisations. Although these are all slightly different coordinate reference systems, this will only be noticeable for applications of the very highest accuracy.

The alignment of WGS 84 with the ITRF means that at the highest level of accuracy it will not be exactly compatible with coordinate reference systems tied to a particular tectonic plate such as ETRS89. WGS 84 differs from ETRS89 by approximately 50 cm (2007 figure), increasing at about 2.5 cm per year. For sub-metre accuracy applications, this is significant, but for applications with less demanding accuracy requirements the difference may be ignored, and WGS 84 coordinates accepted as ETRS89 coordinates, or vice-versa.

WGS 84 is utilised specifically by the GPS system. Other global navigation satellite systems such as the Russian Glonass system and the European Galileo system operate in similar realisations of the ITRF – for Glonass it is the PZ90 coordinate reference system. For many practical purposes, these other GNSS systems can be assumed to deliver WGS 84 coordinates.

2.3.2.5 Why latitude and longitude are not unique

Referring back to section 2.2.2 and Figure 2.5, it will be recalled that, for an ellipsoid, latitude is the angle measured at the intersection of the ellipsoidal normal and the equatorial plane. Figure 2.12 shows a section through the ellipsoids used by two different geodetic datums. It can be seen by inspection of the figure that the latitude of the point P changes when the geodetic datum is changed.

The converse of this holds true. One latitude value will describe different positions if referenced to different coordinate reference systems. The same applies to longitude and ellipsoidal height: if the coordinate reference system is changed, the longitude and ellipsoidal height of a point will change. This point is illustrated by Figures 2.13 and 2.14.

Figure 2.13 illustrates the horizontal component. One latitude and longitude pair describes the location of three different buildings, depending upon the geodetic datum.

The horizontal difference between the same coordinate values referenced to different geodetic datums is typically 50 to 500 m, but in extreme cases can exceed 1.5 km. The corollary is that if the geodetic datum is not specified, mislocation may be of this magnitude. Conversely, the identification of geodetic datum is not critical for low accuracy applications: accuracy of no better than approximately 1 km or mapping at scales of 1:500 000 or smaller (1:1 000 000, 1:2 500 000, etc.).

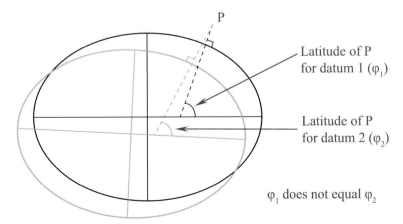

Figure 2.12 Latitude related to different coordinate reference systems.

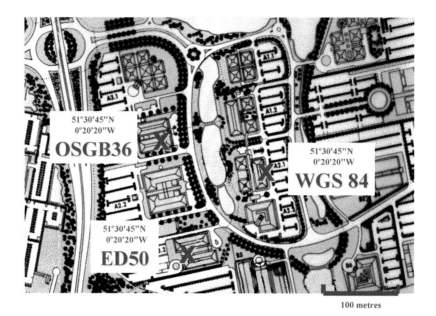

100 metres

Figure 2.13 Latitude and longitude related to different geodetic datums.

Figure 2.14 illustrates the vertical component. It shows the height of the geoid with respect to ellipsoidal heights in the British OSGB 1936 and European ED50 coordinate reference systems, respectively. Note that in these two diagrams the geoid is the same; it is the reference system that has changed. In these illustrations both systems are locally fitted to the geoid; the vertical coordinate values at a point are within a few metres. A globally-fitted system might have a difference in height between the geoid and the geodetic coordinate reference system of ±100 m, as previously discussed in section 2.1.

The general principle here, for horizontal and vertical coordinates, is that coordinates as numbers are ambiguous. They describe location unambiguously only when the associated coordinate reference system is identified.

2.3.3 Projected coordinate reference systems

Map projection coordinates were introduced in section 2.2.5, and are covered extensively in chapter three. What is important to realise at this stage is that the map grid coordinates are derived from ellipsoidal coordinates in a particular geodetic datum and coordinate reference system. The map grid is therefore dependent on the datum. We have seen that latitude and longitude are not unique and are unambiguous only when the datum is identified. This dependence is inherited by map grid coordinates. The Universal Transverse Mercator grid system, discussed in chapter three, may be applied to any geodetic coordinate reference system, as shown in Figure 2.15. UTM coordinates (of any zone) are therefore ambiguous unless the datum to which they are referenced is identified.

Note that in grid coordinate terms the relative positions of ED50 and WGS 84 coordinates differ from those suggested by geographic coordinates shown in Figure 2.13. This is due to the systems using different ellipsoids.

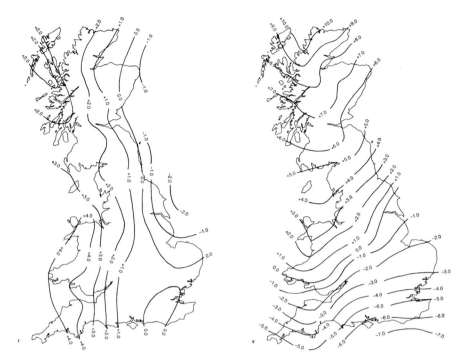

Figure 2.14 The geoid with respect to the British coordinate reference system OSGB 1936 (left) and the old European coordinate reference system ED50 (right). (Courtesy of J. Olliver, Oxford University.)

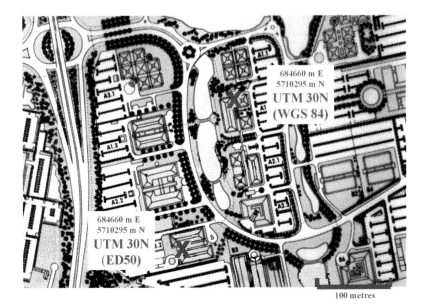

Figure 2.15 UTM coordinates related to different geodetic datums.

2.3.4 Vertical systems
2.3.4.1 Vertical datums and coordinate reference systems
As we have seen in section 2.1, the geoid is defined as that equipotential surface that most closely corresponds to mean sea level. We need to determine mean sea level empirically. This is done by measuring the water level at a tide gauge over a protracted period (typically a minimum of a month and ideally at least 18.9 years). Each of these determinations can provide the origin for a vertical datum. For example, a tide gauge at Qingdao has been used as the origin for heights in China. The first datum (Huang Hai or Yellow Sea 1956) was determined from observations of sea level over two years between 1954 and 1956. This was later superseded by the Huang Hai 1985 vertical datum, which was based on 30 years continuous measurements at the same gauge.

In Great Britain, the datum adopted by the Ordnance Survey is the mean sea level measured at Newlyn between 1915 and 1921. This was adopted in 1921, superseding an earlier datum based on sea level measurements at Liverpool during the early 1840s.

There is an additional requirement to promulgate the vertical datum across the area of interest; that is, to *realise* the datum through the adopted vertical coordinate system. This process involves transferring the height at the datum origin to other fixed points (bench marks) across the area through survey observations and adjustments. The resulting vertical coordinate reference system then is not just the definition of the datum origin point, but also its realisation through a set of accepted benchmark values. Due to the errors introduced in realisation, the vertical coordinate reference system is not exactly equipotential.

Were there to be a later series of survey observations and adjustments based on the same origin tide gauge, these would give rise to a second set of benchmark values. This would be a second realisation of the datum and, with either the same coordinate system or a new one, a second vertical coordinate reference system. Both vertical coordinate reference systems may share the same datum origin. In Great Britain there have been three realisations. The first was applied to the Liverpool origin. The second, observed between 1912 and 1921, and the third measured between 1951 and 1956 were both applied to the 1921 determination of sea level at Newlyn. If heights referenced to one realisation are to be compared with those from another, allowance for the difference between the realisations should be made. Another example is the datum used for mainland France, which has its origin as the mean water level measured between February 2nd 1885 and January 1st 1897 at the tide gauge in Marseille. The datum realisation is known as the Nivellement General de la France (French General Levelling Network). Several adjustments of the network have been made, in particular that made under the supervision of Lallemand in the early part of the 20th Century and the readjustment by the Institut Géographique National (IGN) based on reobservations made between 1962 and 1969. Height differences between the Lallemand realisation and the IGN69 realisation vary by up to two decimetres. Administrative districts have differed in their approach to adoption of the new system, resulting in neighbouring communes using Lallemand and IGN69 heights (CNIG 1996).

The adoption of mean sea level at a single location may be unsatisfactory, particularly over large areas, since the errors that accumulate in the geodetic levelling are likely to be greater than any changes in mean sea level with respect to the geoid. For the Australian Height Datum (AHD), the mean sea level at 30 tide gauges around the

Figure 2.16 Locations of tide gauges used for the Danish Height
Datum DVR90 (Schmidt 2000).

coast of continental Australia was adopted (Roelse 1975). For Denmark, 10 gauges
are used (Schmidt 2000), as shown in Figure 2.16. It is clear in these cases that the
height datum is not a true equipotential surface, since each mean sea level measurement
will be slightly different with respect to the geoid.

2.3.4.2 Distinction between ellipsoidal and gravity-related heights

We have seen that heights can be measured above the surface of the ellipsoid as the
vertical component of a three-dimensional ellipsoidal coordinate system (h), or from the
geoid as a one-dimensional vertical coordinate system (H). In section 2.1, we compared
the ellipsoid and the geoid and saw that the differences between the two surfaces were
up to ±100 m. Whilst small compared with the overall size of the Earth, this is certainly
significant in terms of the differences in heights compared to the two surfaces.

The height of the geoid above the ellipsoid is known as the *geoid-ellipsoid
separation*, or often just the *separation*, and is usually given the symbol N. This may
be a positive or a negative quantity, as shown in Figure 2.17.

Some texts refer not to the height of the geoid above the ellipsoid, but to the
undulation of the geoid. This is a rather unsatisfactory term, as the derivation of
the word implies that the phenomenon under consideration is a wave, rather than
a height. We will call N the separation, and will reserve the expression *undulations
of the geoid* to refer to the presence of 'waves' in the part of the geoid in question.
(A geoid without undulations, for example, would imply that at the accuracy required

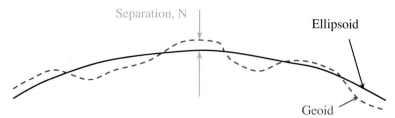

Figure 2.17 Sectional view of an ellipsoid and the geoid.

it appears to be at a constant height above the ellipsoid, or can be modelled simply as an inclined plane.)

Note also in Figure 2.17 that the direction of the vertical (perpendicular to the geoid) is not usually coincident with an ellipsoidal normal. The angle between the two is termed the *deviation of the vertical*, and is usually resolved into a north-south component, ξ, and an east-west component, η. These angles typically amount to a few seconds of arc.

The relationship between the gravity-related height, H, and the ellipsoidal height, h, is as shown in Figure 2.18.

Since the angle between the vertical and the ellipsoid normal is so small, it can be stated without any significant loss of accuracy that:

$$h = H + N \qquad (2.15)$$

It is emphasised that the ellipsoidal height h is the value that normally results from satellite observations; the significance of this equation is therefore that in order to obtain a height above the geoid, H, it is necessary to know the separation, N. *Separation models* that provide this through harmonic expansions or grid interpolation techniques are discussed in section 4.7.

As a final comment here on reference surfaces for height, it is worth pointing out that the advantages of using ellipsoidal heights should not be overlooked. Although these are not always appropriate, there are many situations in which the physical link with the direction of gravity, and with a surface perpendicular to this, is unnecessary. An example of this might be the monitoring of the movement of a point over time, perhaps one near the summit of a volcano to give an indication of future eruption. In this situation, a *change* in the height is all that is required, and a change in an ellipsoidal

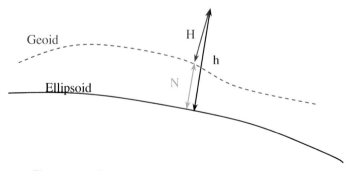

Figure 2.18 Gravity-related height and ellipsoidal height.

height is the same as the change in the gravity-related height to any accuracy that could conceivably be required. The need for a determination of the geoid-ellipsoid separation is thus avoided, as is the danger of an apparent change being recorded as a result of an improved determination of the separation at a later date. To take another example, an aircraft landing at an airfield and using a satellite positioning system (giving ellipsoidal height) can determine its height above obstructions that have themselves been surveyed using satellite techniques and have their heights expressed in ellipsoidal terms in the same datum. In this context, the position of the mean sea level surface becomes an abstract and irrelevant concept.

What might be thought of as a half-way position between using gravity-related heights and ellipsoidal heights is to adopt a standard model of the geoid to correct satellite observations: if the model is not perfect then true gravity-related heights will not be obtained. On the other hand, if the use of the model is universal then only the most precise engineering applications will need to use any other form of height coordinate, and the advantage will be that a consistent system will be adopted. Height correction models are discussed in section 4.7.

2.3.4.3 Vertical datums for marine applications

For nautical charting purposes, the critical information required is the depth of water available for under-keel clearance. In tidal areas, the use of mean sea level (MSL) as the datum is inappropriate because the depth available may be less than that below MSL. Instead a water level below which the tide does not normally fall is chosen as the reference for depths on nautical charts – *chart datum* (CD). In conducting surveys for the preparation of nautical charts, hydrographic surveyors correct for the actual tidal level at the time of a depth measurement and relate all measured depths to a common datum – *sounding datum*.

Tides are complex phenomena, induced primarily through the gravitational attraction of the moon and sun and in shallow water significantly modified by marine topography, but also perturbed by resonance and meteorological effects such as atmospheric pressure and wind. The principal systematic factors may be described through a series of harmonic constituents. The summation of these constituents defines highest and lowest astronomic tide.

The International Hydrographic Organisation (IHO) recommends that where tides have an appreciable effect on water level, lowest astronomic tide (LAT) be adopted as chart datum (IHB 2005). This resolution post-dates practices in many countries and in some of these other water levels have been adopted as chart datum. For example, in coastal waters of the United States mean lower low water (MLLW) is used for both sounding datum and chart datum. In the United Kingdom, LAT was in use for chart datum before the IHO resolution. In maritime areas where there is little or no tide or a complex tidal regime, other water levels may be used.

Chart datum is defined locally. Figure 2.19 shows several discrete points at which CD has been defined. Each definition may not exactly agree with the chosen water level due to small errors in the measurements used – chart datum therefore can be considered to be a realisation of LAT.

At this stage it needs to be emphasised that the *aim* is to make the maritime datum equal to LAT, but once the realisation has been carried out from tidal observations the level is selected and referenced to bench marks onshore and then becomes, in its

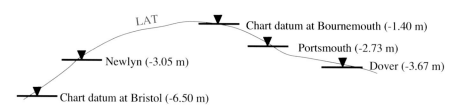

Figure 2.19 Chart Datum relative to the onshore vertical datum and Lowest Astronomic Tide at selected south of England ports.

own right, *Chart Datum*. That is, any subsequent observations that determine that the tidal regime has changed slightly or that the original observations were inaccurate are essentially irrelevant. These will change the position of LAT with respect to Chart Datum but they will not change the datum.

Both tidal and non-tidal water level datums may be referred to the local onshore vertical datum through levelling measurements. However, vertical coordinate reference systems based on water level datums differ from systems used for onshore applications in two distinct ways. Firstly, and most obviously, the coordinate system uses depths rather than heights. Secondly, the water level datum is not an equipotential surface. A water level datum such as LAT varies from mean sea level by amounts that depend upon the tidal range at that point. Largely due to the effects of marine and coastal topography, the tidal range varies from place to place.

A water level such as LAT does not follow an equipotential surface. But it is a continuous surface. For example, along the south coast of England LAT differs from Ordnance Datum Newlyn (ODN) by 3.05 m at Newlyn itself, but by 3.67 m at Dover, 2.73 m at Portsmouth, 1.40 m at Bournemouth, and 6.50 m at Bristol (Avonmouth) (POL 2007). Figure 2.19 shows this schematically, but it should be remembered that a vertical datum such as ODN is not itself an equipotential surface but a realisation of one.

It is relatively straightforward to determine these levels at tide gauges. In between gauges, or out to sea, the interpolation or extrapolation of levels is more problematic. This would usually be accomplished by assuming a smooth interpolation or by finding more information on the change in the tidal regime through such means as *co-tidal* and *co-range* charts. An alternative is to make additional tidal observations in the open sea by dropping a sea-bed pressure sensor but, of course, this cannot be related to the land datum.

For offshore engineering purposes, adopting a water level such as LAT or its realisation through Chart Datum as vertical datum is inappropriate. To determine correct relative depths of the seafloor topography, a project vertical datum that has a fixed offset from an equipotential surface (approximately MSL) must be used. This is particularly the case in areas across which there is significant change of tidal range.

There is one final datum to consider in the marine context, and this is the one that is used to express *heights* of objects *above* the water, the most obvious example being the clearance height of a bridge over a navigable river. In this instance, we are interested not in the safe depth under the keel but in the safe height above the mast or superstructure, and therefore require a conservatively *high* datum. In the United Kingdom, for example, the level used is *mean high water springs* (MHWS). Again, its position with respect to a land datum varies with the tidal regime. It will be noted that it is not such a conservatively high level as highest astronomical tide, presumably because – unlike for depth applications – the mariner usually has the advantage of being able to see the hazard.

Chapter 4 discusses the models that are available for transforming between marine and land datums and coordinate reference systems.

2.3.5 Engineering datums and coordinate reference systems

Coordinate systems for engineering survey applications were introduced in sections 1.2 and 2.2.6. They are used either on Earth-fixed bodies such as construction sites, buildings, or permanent offshore installations, or they are used on moving platforms such as vehicles, vessels, or aircraft.

The *datum* element concerns the way in which these are attached to the body being measured to form an engineering coordinate reference system. For an Earth-fixed system, usually associated with projects in which conventional survey equipment will be used, there would generally be a physical mark that is designated as the 'datum point'. This would have assigned to it coordinates that are sufficiently large as to avoid the appearance of negative values anywhere within the project area. It may also have a height value assigned to it, either loosely related to actual height above sea level or arbitrary, and of sufficient size if this is not the lowest point. A second point, visible from the first, would be designated as a reference object, and assigned either a bearing of zero or some other figure usually to make it correspond to the approximate direction of true north. Either way, what has been established is a direction of 'plant north'. In some cases, plant north may be aligned with one side of the site and differ markedly from geographic north. Although many engineering coordinate reference systems will use a two- or three-dimensional Cartesian coordinate system, special applications will require use of the more rarely encountered coordinate system types – affine, polar, cylindrical, spherical or linear – described in section 2.2.8.

On a moving platform (or one that is capable of movement), the datum origin may be a physical point on the object as with Earth-fixed systems. Alternatively, it may be a virtual point formed at the intersection of the coordinate system axes, which are themselves aligned with some of the features of the platform.

If it is required to relate the coordinate reference system on a moving platform to a global coordinate reference system such as WGS 84, this is usually achieved by on-board processors that take as input position observations with GPS and orientations from an inertial measurement system. Chapter 4 covers the techniques that are used to relate Earth-fixed engineering coordinate reference systems to geodetic and projected reference systems.

2.3.6 Image datums and coordinate reference systems

An image coordinate reference system is a type of engineering coordinate reference system, in that it describes positions with respect to a particular object rather than to the Earth itself. In this case, the object is the photo-sensitive plane of a camera or other sensor, or its reproduction in digital form on a computer. However, an image system is distinguished by its two-dimensional Cartesian or affine coordinate system being presented as an array of adjacent cells. Position within the array can be presented as 'cell units' (counting left/right and up/down from an origin).

An essential part of the image datum information, and the attribute that differentiates this class of datum from ordinary engineering datums, is the specification of the way that the image array is associated with the data attributes of the image. The image grid may have the grid lines running through the centres of the cells, in which case all cell centres will have integer grid units, or the image grid will be associated with the cell corners. This difference in definition has no effect on image interpretation but it is essential information for transforming image coordinates.

The image datum definition applies regardless of whether or not the image is geo-referenced. Geo-referencing is performed through a (sometimes complex) transformation. The image coordinate reference system definition is independent of the transformation.

2.4   Compound coordinate reference systems

There is often a need to describe positions in three dimensions using easting, northing and gravity-related height. Examples include horizontal coordinates in British National Grid terms and gravity-related height referenced to Ordnance Datum Newlyn, or in Australia to describe location using Map Grid of Australia coordinates and AHD height. Alternatively, a data set for Canada and the United States may include latitude and longitude with respect to the geodetic coordinate reference system NAD83 and gravity-related height with respect to the NAVD88 vertical coordinate reference system. In all of these cases, two independent coordinate reference systems are being used, one for horizontal position and one for vertical position.

In coordinate reference system data modelling, these hybrid coordinate sets are described as *compound coordinate reference systems*.

Typical combinations of systems are:

- A projected CRS with a vertical CRS, as in the British and Australian examples above.
- A geographic 2D CRS with a vertical CRS, as in the North American example above.
- An engineering 2D horizontal CRS with a vertical CRS. This would be as used on a construction site where the an arbitrary 'plant grid' was used, but heights were levelled in from local bench marks because there was a requirement for services such as drains to link at the correct level with neighbouring areas.

A compound coordinate reference system is a construct for three-dimensional positioning, and as such it is not a vehicle for generally combining multiple systems. For example, it cannot contain a three-dimensional geodetic system and a gravity-related vertical system.

The attributes of a compound coordinate reference system are the names of its horizontal and vertical systems. Each of these is described as discussed in the earlier sections of this chapter.

The relationship between these compound three-dimensional systems and true three-dimensional coordinate reference systems is discussed in section 4.7.3.

2.5   Coordinate reference system identification

2.5.1 CRS description

For coordinates to identify position unambiguously, the coordinate reference system needs to be fully described. This requires all of the critical attributes of the coordinate system and datum that have been discussed in the previous sections of this chapter, and summarised in Table 2.2.

Note that the definition of the datum is not normally required, the name being sufficient to identify it unambiguously. However, it may sometimes be necessary or useful to include other normally non-essential information. For example, by default 'Ordnance Datum, Newlyn' would be presumed to be the latest realisation. If for some reason it were necessary to reference an earlier realisation with the same name then this should be explicitly stated. Similarly for engineering coordinate reference systems, an explicit description of the datum may be appropriate.

In principle the coordinates for each point must have their reference system described. However, data is normally distributed in sets related to the same system. This being the case, one reference system description can be applied to the whole set. But if any of the essential defining attributes are changed, then a new coordinate reference system description is required.

Table 2.2 Essential information required for coordinate reference system definition.

Attribute	Coordinate reference system type				
	Geodetic CRS	Projected CRS	Vertical CRS	Engineering CRS	Image CRS
CRS name or identifier	✓	✓	✓	✓	✓
Datum name	Note 1	Note 1	Note 1	Note 1	Note 1
Grid relation to cell					✓
Ellipsoid name	✓	✓			
Ellipsoid parameter values (note 4)	✓	✓			
Prime meridian name	Note 2	Note 2			
Prime meridian Greenwich longitude	Note 2	Note 2			
Map projection name		✓			
Map projection method		✓			
Map projection parameter values (note 4)		✓			
Coordinate system definition	✓ (note 3)	✓ (note 3)	✓ (note 3)	✓ (note 3)	✓ (note 3)

Notes: (1) normally omitted if CRS name is datum name.
(2) normally omitted if name is Greenwich and Greenwich longitude is zero.
(3) for each axis includes axis order, direction and units.
(4) including units.

2.5.2 Registers of coordinate reference systems

Rather than provide directly all of the information required for coordinate reference system description, it may be more convenient to identify a coordinate reference system indirectly through reference to a repository of definitions. Examples include the Information and Service System for European Coordinate Reference Systems (BKG 2007) and the global EPSG[5] Geodetic Parameter Dataset (OGP 2007a). These registries include all of the information required to meet the ISO 19111 standard (ISO 2007).

Each coordinate reference system is given an identifier, which is unique within the registry. For example, in the EPSG dataset the unique identifier for the NAD83 Canadian Spatial Reference System is the numeric code 4954 for the geocentric form of the coordinate reference system, code 4955 for the three-dimensional geographic system, code 4617 for the two-dimensional geographic system, and code 2955 for the projected coordinate reference system including UTM zone 11N. It then becomes possible to identify a coordinate reference system by reference to the register and its unique code. For example, easting and northing coordinates referenced to NAD83(CSRS) / UTM zone 11N may be identified by reference to the EPSG dataset as EPSG:2955. If the registry happens to be online and offering a computer service interface, as is the EPSG dataset, the reference may cite a Uniform Resource Identifier, for example urn:x-ogc:def:crs:EPSG:2955. This allows a computer to interrogate the registry and obtain reference details.

Care needs to be taken to ensure that the registry definition exactly matches that required. In the EPSG dataset, the projected coordinate reference system ETRS89 / UTM zone 33N with axes in the order easting, northing has the code 25833, but with axes in the order northing, easting (as sometimes used in eastern Europe) the code is EPSG:3045. Similarly for State Plane systems in the United States, different definitions and codes will be used depending upon whether coordinates are in metres or feet. If the registry definition does not exactly match what is required, then its code should not be used. Instead an explicit definition should be provided.

These registries have internal rules for the data that they will include. For example, the EuroGeographics register includes data for Europe used for national mapping. Some coordinate reference systems may not appear in any register. It is then not possible to use a registry as an indirect identifier of coordinate reference system and it will be necessary to use the explicit description discussed in the previous section.

Many repositories do not delete erroneous data. This is to allow users to refer to the information for historic purposes. In these cases the repository will retain erroneous records, mark them as being invalid, and add a corrected replacement. Unless there is a need to rework coordinates that have already been subjected to an invalid conversion, users should take care to use only valid registry data.

Some of these coordinate reference system registers may also include information on transformations between reference systems. This is discussed further in section 4.8.

---

[5] The dataset was initiated by the European Petroleum Survey Group. This group disbanded and reformed as the Surveying and Positioning committee of the International Association of Oil and Gas Producers (OGP). The abbreviation EPSG has been retained only as an identifier for the dataset.

3

MAP PROJECTIONS

Geodetic Datum A	CRS 1 Geocentric X Y Z	CRS 2 Geographic 3D φ λ h	CRS 3 Geographic 2D φ λ	CRS 4 Projected E N

3.1   Introduction

The fundamental coordinate reference system for surveying and mapping is a set of geodetic coordinates related to a particular datum. It is then necessary to consider how to arrange the data so that it can be placed on a flat surface.

There are two reasons for doing this. The first, and most obvious, is presentational. Whether the data is to be shown on a paper map or on a computer screen, it must of necessity be presented in a two-dimensional format.

The second reason for rearranging the geodetic coordinates in two dimensions is computational. Even a simple concept such as the distance between two points becomes complex when expressed in ellipsoidal formulae, and wherever possible it is more desirable to carry out computations in a simple two-dimensional Cartesian system.

A map projection, then, is defined as an ordered system of meridians and parallels on a flat surface. It should be immediately apparent that it is impossible to convert a sphere or an ellipsoid into a flat plane without in some way distorting or cutting it. It then follows that there is no single method for doing this, and hence the proliferation of types of map projection.

Before looking in detail at the different types of projections, the next part of this chapter (section 3.2) is concerned with introducing some of the fundamental concepts of map projections that it is necessary to grasp. Included here are definitions of grids and graticules, and of scale factor, as well as a consideration of the use of spherical or ellipsoidal models of the Earth in the context of projections, the use of different developable surfaces, and the criteria to be considered in designing a projection for a specific purpose. Some fundamental defining parameters have more conveniently been introduced in the context of specific projections, although they have a universal application. For reference, these are all to be found in section 3.3 and can be listed as:

False coordinates and the *projection origin*: discussed with reference to the cylindrical equidistant projection.

Re-scaling of projections: discussed with reference to the Mercator projection.

39

Zoning: discussed with reference to the Transverse Mercator projection.

Convergence, or the angle between grid north and true north, is also introduced with reference to the Transverse Mercator projection.

An example of the development of formulae to convert between geographic and projection coordinates is given in section 3.3.1, with further comments on the formulae required in section 3.7.1. A summary of the information needed to define or identify a projection is given in section 3.7.2.

Sections 3.3 to 3.6 consider many different methods for projecting coordinates: it will be seen that each of these methods requires several defining parameters. It is therefore important to distinguish between:

- a *projection method* (in which may be classed, for example, the Transverse Mercator, the polar stereographic, and so on);
- a *map projection*, which is composed of the method as well as specific values for the method's defining parameters (and in which would be included, for example, the British National Grid, the Universal Transverse Mercator system, and so on); and
- a *projected coordinate reference system*, which introduces the datum and therefore completely relates the given coordinates to the real Earth, as discussed in Chapter 2.

Two final sections complete this chapter on map projections. Section 3.8 is concerned with computations that are carried out using coordinates from a projected coordinate reference system, in which real-world phenomena such as distance and area have been distorted. Section 3.9 gives advice for those wishing to design a projected coordinate system to meet specific objectives.

3.2 Map projections: fundamental concepts

3.2.1 Grids and graticules

Meridians and parallels appearing on the projection have an appearance that is dependent on the type of projection that has been used. In general they will be an arrangement of straight or curved lines as shown for example in Figure 3.1.

This set of parallels and meridians, as seen on the map, is known as the *graticule*. On some maps it may not be shown at all; on others it may be noted around the border of the map and shown in the middle as a set of tick marks (for example, on Ordnance Survey 1:50 000 maps, where blue tick marks show the graticule at 5' intervals).

This graticule does not constitute the basis of a coordinate reference system that is suitable for computational purposes or for placing features on the projection. Instead, a rectangular coordinate reference system known as the *grid* is superimposed on the map. See Figure 3.2.

This grid is given coordinates and, as was discussed in section 2.3.5, these may be labelled x and y, or eastings (E) and northings (N), or some other convention.

3.2.2 Scale factor

Features on the surface of a sphere or an ellipsoid undergo distortions when projected onto a plane. It is necessary to have a precise definition of the amount of distortion that has resulted.

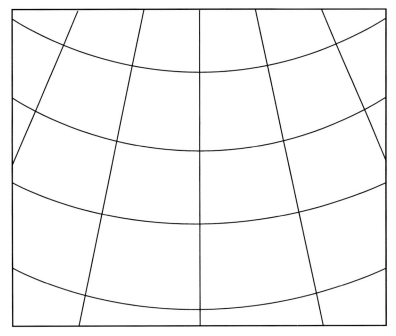

Figure 3.1 **An example of the graticule.**

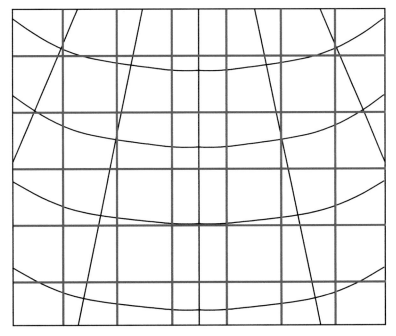

Figure 3.2 **A grid superimposed on the graticule.**

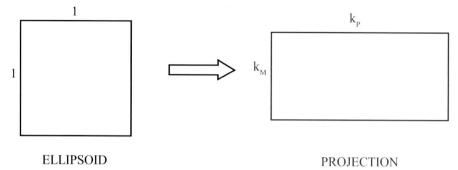

ELLIPSOID   PROJECTION

Figure 3.3 Projection of a unit square from the surface of
the ellipsoid onto the projection.

This is provided by the definition of the *scale factor*, which is given the symbol k in this text. Then:

$$k = \frac{\text{distance on the projection}}{\text{distance on the ellipsoid}} \tag{3.1}$$

This parameter will be different at each point on the projection, and in many cases will have different values in each direction. Equation 3.1 can therefore be understood to apply in general to only a short distance (in theory infinitesimally short). For longer lines, the relevant parameter is the integrated mean of the point scale factors along the whole length of the line: section 3.8 discusses the point at which a short line becomes a long one.

It is important to understand that this scale factor results purely from the act of projecting to a flat surface and is therefore unrelated to the scale of map (a ratio such as 1:50 000). The ideal value of the scale factor is 1, representing no distortion. It should also be emphasised that a distortion of this type is not the same as an 'error' in the map, as the rules governing it are clearly defined, and the true coordinates can always be recovered if the values of the parameters of the projection are known.

It is useful to consider what happens to a small square of dimension (1×1) on the surface of the ellipsoid when it is projected. In the general case, the distortion in the direction of the parallels will be different to the distortion in the direction of the meridians. Let k_p represent the scale factor along a parallel, and k_M represent the scale factor along a meridian. With reference to Figure 3.3, the square is then projected as a rectangle of dimensions ($k_p \times k_M$).

It will be seen in the examples of projections that follow that the unit square is often subjected to a rotation as well.

3.2.3 Developable surfaces

Historically, projections were derived by first projecting from the ellipsoid (or a sphere for a simplified model) to an intermediate surface that was of such a nature that it could be unravelled without distortion. This remains a useful concept for categorising and describing map projections.

The principal forms of these intermediate surfaces are the cone, the cylinder, and the plane itself. The advantage of these shapes is that because their curvature is in one dimension only, they can be unravelled to a plane without any further distortion – the surfaces are *developable*.

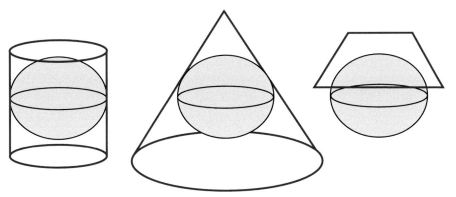

Figure 3.4 **Developable surfaces: (a) Cylindrical; (b) Conic; (c) Plane.**

The developable surface is brought into contact with the ellipsoid. Examples are shown in Figure 3.4.

A set of rules is formulated for the way in which features are taken from the ellipsoid onto that surface. The rules for transferring features from the ellipsoid to the projection are discussed later in this chapter. Before looking at these in detail, however, the general point can be made that in the region around the point or line of contact between the two surfaces the scale factor distortion will be minimal. In fact, where the two surfaces are touching, the scale factor will be equal to 1.

The choice of developable surface will therefore be dictated by the geographical extent of the region to be mapped. In some cases, it is required to project the whole Earth; in others, however, a projection will only apply to a selected area. Thus, for example, a cylindrical surface as shown in Figure 3.4 (a) will be appropriate for mapping the equatorial regions, as it is touching the ellipsoid along the equator. A conic projection as in Figure 3.4 (b) will be good for mapping areas of mid-latitude with a large extent in longitude, as the cone is in contact with the ellipsoid along a line of latitude.

The characteristics of the developable surfaces have only been given in outline here: it is, for example, possible to orientate the surface at different angles to those shown here, or to change the shape of the cone. These will be dealt with more fully in the sections that follow. When the axis of the cylinder or cone is coincident with the polar axis of the ellipsoid, the aspect of the projection is referred to as 'normal'.

Cones, cylinders, and planes are helpful for gaining an insight into the appearance of a projection and for understanding the geometry of the projection and how it distorts features such as length and area.

It should be noted here, however, that a developable surface is not a *necessary* step in forming a projection. It is possible to derive a set of formulae to convert geographic coordinates to grid coordinates in purely mathematical terms. In general, equations can be derived of the form:

$$(E,N) = f(\varphi, \lambda) \tag{3.2}$$

which express the grid coordinates as a function of the geographic coordinates without reference to intermediate developable surfaces. Indeed, the equations for all the above projections could be given in this form without mention of cones and so forth, and in this case expressions for aspects such as scale factor and convergence could be derived by differentiation.

Most of the more complex projections, which depart from the simple forms described in the following sections, are usually developed to represent the whole Earth in some way, and so are of less relevance to surveying, remote sensing, and GIS. There are exceptions to this, however, and some of these are discussed in section 3.6.

3.2.4 Preserved features

Having selected the developable surface, it is necessary to devise a set of rules for transferring coordinates from the sphere. In theory, there is an infinite number of ways of doing this, and the choice will depend on the purpose for which the projection is devised.

It is not possible to devise a projection without introducing distortions. In general, the shape, area, and size of features on the surface of the sphere or ellipsoid will be different when converted to the projection. The usual approach is to attempt to preserve *one* of these, which will usually be done at the expense of all the others.

For example, it may be required that certain of the distances as measured on the sphere should be undistorted when shown on the projection. It is obviously not possible to preserve *all* distances, as this would then be achieving the unachievable goal of an undistorted projection. It may be instead that the distances along all meridians should remain undistorted, which is the same as saying that:

$$k_M = 1 \qquad (3.3)$$

or the scale factor along a meridian is equal to one. The effect on the projection of a unit square is then shown in Figure 3.5.

Such a projection is termed an *equidistant* projection. It can be seen that there remains a scale factor along the parallels, which is not equal to one, and that both the shape and the area of the square have been distorted.

An alternative to this type of projection is one that attempts to preserve area, and is therefore termed an *equal area projection*. In such a situation it is clearly the case that:

$$k_M \, k_P = 1 \qquad (3.4)$$

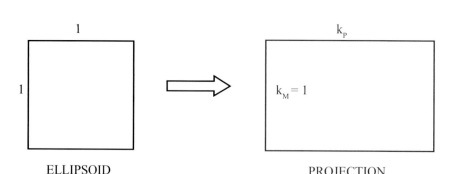

ELLIPSOID   PROJECTION

Figure 3.5 Projection of a unit square preserving distances along the meridians.

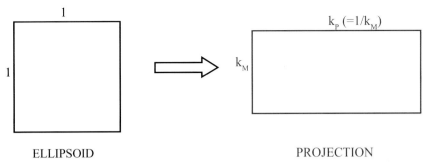

ELLIPSOID   PROJECTION

Figure 3.6 **Projection of a unit square preserving area.**

or that the area of the projected unit square remains equal to one. This is illustrated in Figure 3.6.

The other principal classification of projections is that which preserves the *shape* of features. This is known as an *orthomorphic* or, more commonly, *conformal* projection and the relationship between the scale factors is:

$$k_M = k_P \qquad (3.5)$$

This is illustrated in Figure 3.7.

In preserving shape, a conformal projection is therefore preserving *angles* as well. For example, the angle between the side of the unit square and the diagonal is 45°: this is the angle that would be measured by someone making the observation on the ground. With reference to Figure 3.7, the angle between the side and the diagonal is also 45° for the projection. This is not the case, however, in either Figure 3.6 (the equal area projection) or Figure 3.5 (the equidistant projection).

For this reason, the conformal projection is the one of most significance in land surveying, as it means that angles measured on the ground can be transferred to the projection for use in computations. A conformal projection is therefore one of the most frequently used, and is likely to be the basis of almost all large scale mapping.

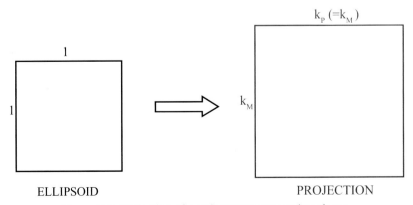

ELLIPSOID   PROJECTION

Figure 3.7 **Projection of a unit square preserving shape.**

Finally, it should be noted that the three types of projection mentioned above, whilst being the most commonly used of all projections, do not constitute an exhaustive list. Other types that preserve neither shape, area, nor any distances are possible, and are sometimes used. Some of these will be referred to in later sections.

It should be emphasised here that the unit square used in these examples has to be a very small one for the conclusions drawn here to be exact: for bodies of finite size, some of these assumptions can break down. This point is illustrated in Case Study 6.4.

In summary, most projection methods may be classified firstly according to the shape of the developable surface, which is dictated primarily by the geographical area to be mapped but also in part by the function of the map, and secondly by the features on the sphere that are to be preserved on the projection.

3.2.5 Spheres and ellipsoids

As mentioned in the introduction to this chapter, the fundamental coordinate system is an ellipsoidal one related to a geodetic datum using an ellipsoid. The relative positions of points in such a coordinate system are different to how they would be on a sphere. That is to say, if accuracy is to be preserved it is necessary to develop formulae for treating an ellipsoid rather than a sphere.

That said, it should be borne in mind that the flattening of most ellipsoids is in the order of one part in 300. It is therefore apparent that the shape of a particular projection when applied to the sphere is very similar to the shape when it is applied to the ellipsoid. There are significant differences in the coordinates that result, which will certainly be apparent on a large scale map, but the sphere is nevertheless useful for giving an insight into how the resulting map has been distorted.

In the discussions of individual projection types in the sections that follow, diagrams are used in which the figure of the Earth is indistinguishable between a sphere and an ellipsoid, and formulae derived from a spherical model are sometimes used to give approximate indications of the amount of scale factor distortion and so on. The reader should note, however, that full ellipsoidal formulae should normally be used in practice. Before moving on from this topic, we shall discuss some of the consequences of confusing the two models.

When merging datasets – for example when overlaying your own data on maps or satellite imagery from another source – it is critical that all datasets are referenced to the same coordinate reference system.

For example, despite the implication that they use the WGS 84 system, mapping applications such as Microsoft Virtual Earth™ and Google Maps™ use a spherical development of the Mercator projection (see section 3.3.3). This is a perfectly good assumption when mapping large parts of the Earth at small scales, but the approximation begins to break down as one zooms in.

Figure 3.8 shows the difference in the length of a degree when using a spherical earth model with a radius R of 6 378 137 m, compared to the length of a degree using the WGS 84 ellipsoid in which the semi-major axis (a) is 6 378 137 m and the ellipsoid is flattened by 1/298.2572236.

Latitude differences are minimised in mid-latitudes, whilst longitude exhibits maximum differences in mid-latitudes. Differences of up to several hundred metres arise. Is this significant? A map of the whole Earth on a 20 inch computer screen may

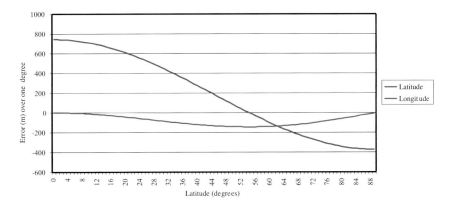

Figure 3.8 Differences in the length of a degree between ellipsoidal and spherical earth models.

be at a scale of 1:20 million. At this scale, 800 m plots as 0.04 mm and cannot be discriminated within one pixel.

As one zooms in, one is more interested in the relative distances between points. The percentage error in scale as shown in Figure 3.9 then becomes more relevant. It does become significant when the difference, which may be considered to be caused by an error in selection of the model of the Earth, is the same magnitude as a pixel. This occurs at a scale of approximately 1:20 000.

When using a Mercator projection, however, northing coordinates are calculated from the equator. It is differences in the distance from the equator that become important. Similarly, using an origin on the Greenwich meridian, for eastings the longitude from Greenwich becomes important. When the distances from the origin are large, data will not overlay properly, and the implication of Figure 3.8 is that the discrepancies

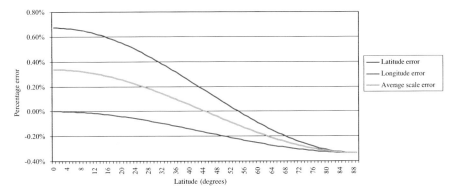

Figure 3.9 Scale error as a function of latitude due to using a spherical earth model.

that would result from mixing coordinates derived from ellipsoidal and spherical models could amount to several hundred metres, or even a few kilometres.

The difference between spherical and ellipsoidal distances could be slightly reduced by taking the radius of the sphere as the recognised value of 6371 rather than 6378 km, but this would still not produce a sufficiently precise match for most applications.

One final comment on the sphere and the ellipsoid is that the latter is not actually a conformable surface. This means that when using the ellipsoid as the model, formulae need to be developed (generally invisibly to the user) that first convert from the ellipsoid to the sphere. As with projections, there is more than one way of doing this and this can lead to slight differences in ellipsoidal projection formulae that have been developed by different people. Sometimes (as with the differences between Transverse Mercator formulae and the Gauss-Krüger method discussed in section 3.3.4) the differences can be ignored for most practical applications. In other circumstances (for example the different forms of the ellipsoidal stereographic formulae discussed in section 3.4.4), the differences can lead to significant problems. To avoid problems, the general rule is: *use the models, formulae, and parameters that have been specified for the coordinate reference system in which you are working.*

3.3 Cylindrical projections

3.3.1 Cylindrical equidistant

Following the rules and procedures outlined in section 3.2, a cylindrical equidistant projection is formed by bringing a cylinder into contact with the sphere and 'peeling' the meridians off the sphere and onto the cylinder without any distortion. This maintains the feature that $k_M = 1$.

In so doing, it is necessary to stretch each parallel of latitude to be the same size as the equator. Then, if the circumference of the equator, L_{eq}, is given by:

$$L_{eq} = 2\pi R \tag{3.6}$$

where R is the radius of the spherical model of the Earth, and the circumference of a parallel of latitude φ is given by:

$$L_\varphi = 2\pi R \cos\varphi \tag{3.7}$$

then by the original definition of scale factor in equation 3.1:

$$k_p = \frac{2\pi R}{2\pi R \cos\varphi} = \sec\varphi \tag{3.8}$$

An example of this projection for the European region is shown in Figure 3.10. The features to be noted are:

1. As with all normal aspect cylindrical projections, the meridians are straight and parallel to each other.

2. The distances along the meridians are undistorted.

3. The scale along the equator is true, but the scale of all other parallels becomes increasingly distorted towards the poles, with the extreme case of the poles themselves being represented as straight lines. Here it should be noted that:

$$\sec 90° = \infty \tag{3.9}$$

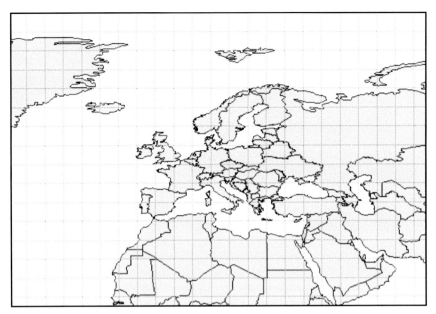

Figure 3.10 **Cylindrical equidistant projection.**

4. As a consequence, the shape and the area become increasingly distorted towards the poles.

5. The flat, square appearance of the graticule leads to the French term *plate carrée*, which is sometimes also used in English for certain forms of this projection.

In computational terms, it is necessary to have a set of formulae to determine the coordinates of points on the map (E, N) given their geographic coordinates (φ, λ).

The first step is to select an *origin for the projection*. Here it may be conveniently chosen, for example, as:

$$\varphi_0 = 30° \qquad \text{(the latitude of origin)}$$

$$\lambda_0 = 0° \qquad \text{(the longitude of origin)}$$

The E (eastings) coordinate is calculated as the distance along the equator between the projected point and the origin, or:

$$E' = R\,\Delta\lambda \tag{3.10}$$

where

$$\Delta\lambda = (\lambda - \lambda_0) \tag{3.11}$$

and is expressed in radians. The expression E' is used at this stage as this formula does not yet represent the finally adopted value of the eastings.

Similarly, the N (northings) coordinate is expressed as the distance along a meridian from the origin to the projected point, or:

$$N' = R\,\Delta\varphi \tag{3.12}$$

where
$$\Delta\varphi = (\varphi - \varphi_O) \tag{3.13}$$

This then has the effect that any points south of 30°N or west of 0° will have negative coordinates on the projection. This is generally undesirable, and is avoided by adding a suitably large number to all the coordinates obtained so far. Thus:

$$E = E' + E_O \tag{3.14}$$

$$N = N' + N_O \tag{3.15}$$

where E and N now represent the *eastings* and *northings* of the point and E_O and N_O are the easting and northing at the true origin – the *false eastings* and *false northings* – which are parameters to be defined for the projection.

To clarify what we have done here, and to introduce some of the terminology that is required in consequence:

- Most projections can be thought of as having a *natural origin* where the ellipsoid and the developable surface are in common. The geographic coordinates of this point are usually parameters that are required to define the projection. For this projection, it would have had to be a point on the equator (0° latitude) with a longitude that we were free to select. Then, to prevent negative coordinates south or west of this point, we assign it *false easting* and *false northing* (FE and FN) values. So if we start with (0, 0) as the coordinates of the natural origin, we need to add E_O and N_O to determine the final eastings and northings.
- In this example, we have chosen a slightly different route, in that we have selected an origin (30° φ, 0° λ) that is *not* a natural one. In this case, we have used a *false origin* and assigned the *easting at the false origin* and the *northing at the false origin*. In this case we have used the symbols E_O and N_O; in other circumstances E_F and N_F may be used.
- The geographic coordinates of the natural or the false origin, and any false eastings and northings applied, are essential defining parameters of the projected reference system.
- In all cases there is a *grid origin* that has coordinates (0, 0) in the final projected system. There is no intrinsic interest in this point, beyond it setting a south-western limit to the applicability of the coordinate system, and we do not use it as a defining parameter.

It is also possible to reverse the computation to derive the geodetic coordinates of any point whose projected coordinates are known, thus:

$$\varphi = \frac{\varphi_O + N - N_O}{R} \tag{3.16}$$

$$\lambda = \frac{\lambda_O + E - E_O}{R} \tag{3.17}$$

Several important points should be noted.

1. These formulae are specific to the cylindrical equidistant projection. They have been written in full as an example of the information that is necessary for defining

a projection: it is not possible in a volume such as this to quote full formulae for all projections. It is to be assumed that the reader will have access to the necessary formulae via software packages commonly used in geomatics and mapping. Further comments on this are given in section 3.7.1.

2. The formulae given above were derived using a spherical Earth. Those for an ellipsoid are more complex.

3. The concepts of the *latitude and longitude of origin* of the projection and the *false easting and northing* coordinates have for convenience been introduced with reference to the cylindrical equidistant projection. It should be noted, however, that these are applicable to *any* projection method, and are a part of the set of parameters that define a particular projection.

4. Equation 3.11 assumes that longitudes are in the range $-180° \leq \lambda \leq 180°$ or $-\pi \leq \lambda \leq \pi$ radians. If the area of interest crosses the 180° meridian, longitudes need to be corrected to fall into a range of $-\pi \leq \lambda \leq \pi$ radians. This may be achieved by applying the following:

$$\text{If } (\lambda - \lambda_O) < -\pi \text{ radians}, \quad \lambda = \lambda + 2\pi \tag{3.18}$$

$$\text{If } (\lambda - \lambda_O) > \pi \text{ radians}, \quad \lambda = \lambda - 2\pi \tag{3.19}$$

Equation 3.18 may be required when $\lambda_O > 0$ and equation 3.19 may be needed when $\lambda_O < 0$. These corrections are not limited to the cylindrical equidistant projection but apply to all projection methods.

3.3.2 Cylindrical equal area

It was shown in the previous section that the cylindrical equidistant projection distorts parallels by the scale factor secφ whilst leaving the meridians undistorted. It is therefore the case that areas have also been distorted.

To compensate for this, an equal area projection can be formed according to the rule of equation 3.4 that $k_M k_P = 1$. Then, since the parallels must be distorted by secφ whatever happens (to fit onto the cylinder), this leads to the conclusion that:

$$k_M = \frac{1}{k_P} = \cos\varphi \tag{3.20}$$

Hence, each small section of each meridian is multiplied by cosφ as it is 'unpeeled' and placed on the projection. This leads to a result as in Figure 3.11.

The features to be noted are:

1. The scale factor in the equatorial region is close to 1 for both the meridians and the parallels. This is a consequence of cylindrical projections being optimal for the equatorial regions.

2. The scale factor along the meridians is no longer equal to 1, and hence distances cannot be measured directly off such a map. Furthermore, the correction to be applied is not a straightforward one, as the scale factor is a function of latitude.

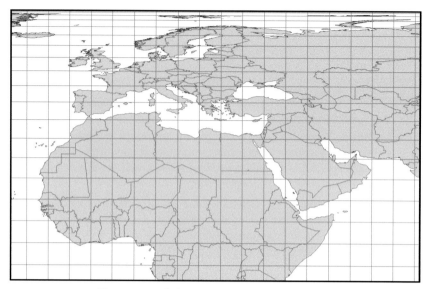

Figure 3.11 Cylindrical equal area projection.

3. The distortion of shape is now extreme towards the poles, as in addition to the scale factor $\sec\varphi$ along the parallels there is the distortion $\cos\varphi$ along the meridians.

4. Formulae can again be derived for converting between (φ, λ) and (E,N), which require the coordinates of the origin and the false coordinates as defining parameters.

A further refinement of this projection is to keep the equal area property but to change the shape, by applying a further scaling of 0.5 along the parallels and 2 along the meridians. This leads to what is usually referred to as the Peters Projection, often used by international organisations for displaying the countries of the world in their correct relative sizes. This is shown in Figure 3.12. It is noted that the shape of

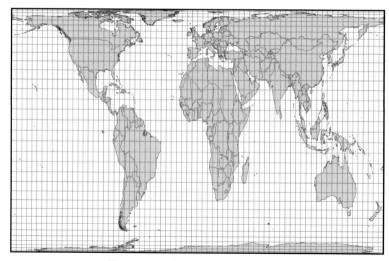

Figure 3.12 The Peters projection.

features is now correct in the mid-latitudes, as opposed to the equatorial regions with the conventional form, and that there is less shape distortion near the poles.

3.3.3 The Mercator projection

One of the most important of all cylindrical projections is the conformal version, which is given the particular name *Mercator*. In this projection, it is again noted that $k_p = \sec\varphi$, and hence from equation 3.5:

$$k_M = k_p = \sec\varphi \qquad (3.21)$$

This then leads to a projection such as that shown in Figure 3.13. The general features of this are:

1. The scale factor at any point and in any direction is equal to $\sec\varphi$, the secant of the latitude.

2. In consequence, the pole is now of infinite size and at infinite distance from the equator, and hence cannot be represented on the projection.

3. The fact that the meridians are parallel to each other and that the angles are preserved makes this an ideal projection for navigation. A line drawn between two points on the map, A and B, as shown in Figure 3.14, has a constant angle with respect to the meridians (the *azimuth from north*), which can be read directly from the map. This is then the azimuth that should be followed in navigating from A to B.

   Such a line is termed a *rhumb line* or a *loxodrome*. It should be noted, however, that this line is not the shortest route between A and B, due to the variation of scale factor within the projection. The shortest route between the two, the *great circle*, in most cases will in fact be projected as a curved line.

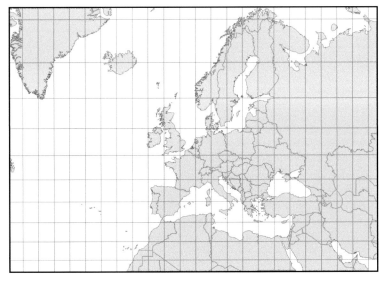

Figure 3.13 **The Mercator projection.**

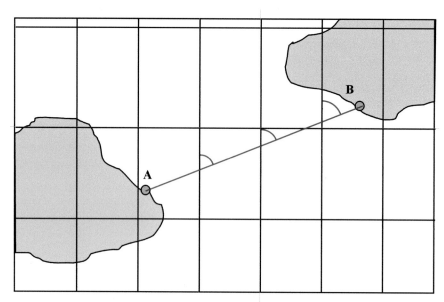

Figure 3.14 **A line of constant azimuth from A to B on a Mercator projection.**

4. An individual map sheet (or navigation chart) will usually represent a small part of the world. It will hence have a scale factor that varies within the map according to the range of latitudes it represents. A chart of the English Channel, for example, might represent the range of latitudes between 49°N and 51°N. The scale factor would then vary between sec49° and sec51°, or 1.52 and 1.59.

It is appropriate then to apply an *overall scaling* to the map so that the scale factor is on average equal or closer to 1. This will not affect the shape of the map in any way, but will make any distances read from it closer to their true values. In this example, an appropriate figure might be

$$k_o = \frac{1}{1.55} = 0.645$$

so that the scale factor is now seen to vary between $k_o$sec49° and $k_o$sec51°, or 0.98 and 1.03, which means that a distance read from the map will be correct to within 3%.

Since the scale on the equator was originally equal to 1, after applying the overall scale factor it is now equal to k_o.

A coordinate conversion or mapping software package usually achieves this by asking the user to input the latitude of the parallel at which the scale is true, also known as the *standard parallel*. If the user has a projection defined in terms of the scale factor on the equator, k_o, the appropriate parallel may be computed by knowing that the general scale factor at any point, k, is now equal to $k_o$secφ.

This concept of an overall *re-scaling* of the projection is, once again, one that has been introduced for a specific example but which has a general application to all map projections. It, or its equivalent, then forms another of the parameters required to define a projection – the *scale factor at the origin*.

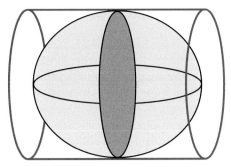

Figure 3.15 **The transverse aspect of the cylindrical projection.**

3.3.4 Transverse Mercator

All of the cylindrical projections discussed so far were formed by placing a cylinder in contact with the equator and, whilst they are often used to portray the Earth as a whole, they are therefore optimised for use in the equatorial regions.

For those parts of the Earth that do not lie close to the equator, an alternative is to turn the cylinder onto its side and make the line of contact a particular *meridian*, as in Figure 3.15.

A projection so formed is termed a *transverse cylindrical* projection, and can be based on any chosen longitude of origin. Again, a set of rules can be proposed to produce equal area, equidistant, or conformal projections. By far the most important of these is the Transverse Mercator projection method, an example of which is shown in Figure 3.16, which has been based on 0° as the longitude of origin, sometimes known as the *central meridian*.

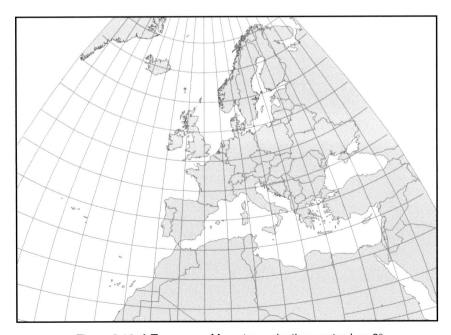

Figure 3.16 **A Transverse Mercator projection centred on 0°.**

The important features of this projection method are:

1. The projection is conformal.
2. The scale factor at each point is the same in any direction, and is given by:

$$k = \sec\theta \qquad (3.22)$$

where θ is exactly analogous to φ, except that it is the angular distance from the central meridian, rather than the angular distance from the equator. It is in a sense a 'sideways version of latitude'. Note that it is not the same as longitude, but for a sphere can be found from the expression:

$$\theta \approx \Delta\lambda \cos\varphi \qquad (3.23)$$

3. The meridians are no longer parallel to each other, and in fact are no longer straight lines. The exception is the central meridian, which is a straight line. At any general point, the meridian makes an angle with the central meridian (which is also the direction of *grid north*) and this angle is termed the *convergence*, γ. For the sphere, it is approximately true that:

$$\gamma \approx \Delta\lambda \sin\varphi \qquad (3.24)$$

Examples of different values for convergence are shown in Figure 3.17.

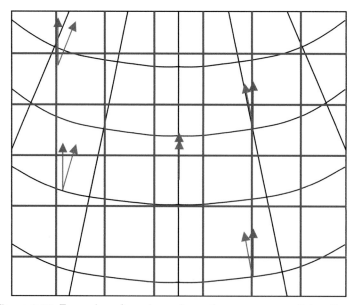

Figure 3.17 **Examples of varying values of convergence within a Transverse Mercator projection. Note that true north is always towards the central meridian, and that the magnitude of convergence increases with distance from the central meridian.**

4. Although Figure 3.16 shows the Transverse Mercator applied to the European region, it is more commonly used for a narrow band of no more than ±3° on either side of the central meridian. This is principally due to its use as a survey projection, in which it is required to minimise the scale factor distortion at the expense of the extent of coverage. In this situation, the scale factor varies between 1 on the central meridian and 1.0014 on the edge of the projection, as shown in Figure 3.18. It is then possible to minimise the scale factor distortion by once again applying an overall scaling k_0, which in the context of the Transverse Mercator projection method is now termed the *scale factor at natural origin*, or, because in this case it applies along all of the chosen meridian through the projection origin, the *central meridian scale factor*. A typical value for medium scale topographic mapping would be 0.9996, which means that the scale factor across the projection now ranges from 0.9996 on the central meridian to 1.0010 (0.9996sec3°) on the edge. This is represented by the red line in Figure 3.18.

5. When, as is the usual case, an ellipsoid rather than a sphere is used as the model of the Earth, the formulae for converting ellipsoidal coordinates to grid coordinates become more complex. The complexity arises because the surface of the ellipsoid is not conformable. Ellipsoidal coordinates are first converted to spherical coordinates before being further converted into Cartesian grid coordinates. The ellipsoidal to spherical conversion involves integration, and several algebraic formulae have been developed to approximate the integral. Gauss-Krüger is one such development. Gauss-Boaga, adopted in Italy, is another. As these approximations are all good – better than 1 mm within three degrees of the central meridian – for all practical purposes the ellipsoidal Transverse Mercator, Gauss-Krüger and Gauss-Boaga methods can be considered to be the same.

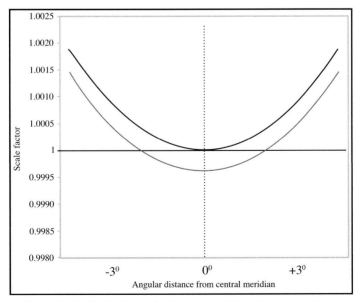

Figure 3.18 Scale factor for Transverse Mercator projections.

The Transverse Mercator projection method is very widely used, and is particularly appropriate for regions with a large extent north-south but little extent east-west. It is, for example, the projection used for the British National Grid (Ordnance Survey 2006) and for several of the State Plane zones in the United States (Stem 1990). Examples of the relevant parameters for selected projections are given in Table 3.1.

We noted above that for large scale topographic mapping purposes, where it is important to keep the scale close to 1, it is usual to restrict the longitudinal extent of a Transverse Mercator projection. The question arises as to what might be done if the area of interest is wider than this limit. One solution is to create two or more similar zones with differing longitudes of origin. The boundary between the zones falls where scale distortion reaches a value that is the maximum acceptable for the purpose intended, as shown in Figure 3.19. The zone boundary is midway between the central meridians of the two zones and on the boundary the scale for each zone is identical.

To distinguish between coordinates of the two zones, a different false easting may sometimes also be defined. Examples of this approach are found in Italy and some States in the US (Table 3.2). Regions that fall across TM zones would be referenced to separate projected coordinate reference systems, with a discontinuity between. Features falling across the boundary would have approximately the same scale factor on either side, but would be rotated with respect to each half since the convergence is in opposite directions (true north is always *towards* the central meridian). This is illustrated in Figure 3.20.

The Transverse Mercator zoning concept may be extended to the whole Earth. It is the basis of a world-wide projection system known as Universal Transverse Mercator, or UTM. This system divides the world up into 60 zones of longitude, each of width 6°. The zones are numbered from 1 starting at a longitude of 180°, and increase eastwards, as shown in Figure 3.21. Thus the UK, for example, lies in UTM zones 30 and 31. Each zone is divided into two, for northern and southern hemispheres. This is to ensure that coordinates in the southern hemisphere can be made positive through the introduction of a false northing.

Table 3.1 Projection parameters and their values for selected Transverse Mercator projections.

Projection	Projection parameter values				
	Latitude of origin φ_0	Longitude of origin λ_0	Scale factor at origin k_0	False easting FE	False northing FN
British National Grid	49°N	2°W	0.9996012717	400 000.00 m	−100 000.00 m
New Zealand Transverse Mercator	0°N	173°E	0.9996	1 600 000.00 m	10 000 000.00 m
New Jersey State Plane CS83	38°50'N	74°30'W	0.9999	150 000.00 m 492 125.00 ftUS	0.00 m 0.00 ftUS

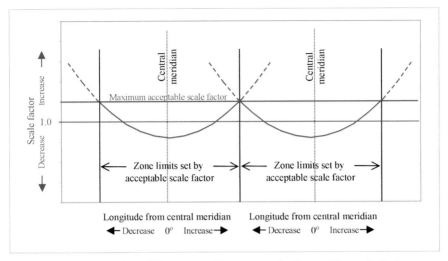

Figure 3.19 **Zoning of Transverse Mercator projection set by scale factor.**

Within each zone, the longitude of origin lies on the meridian along the centre of the zone. Thus, UTM zone 1 has central meridian 177°W, for UTM zone 2 it is 171°W, and so on. All other parameters of the UTM system are as given in Table 3.3.

UTM is not the only global zoned Transverse Mercator system. In Russia, China and much of continental Europe, the Gauss-Krüger zoned system is used. Confusingly, the Gauss-Krüger method name has also been adopted as the name of the zoned system. There are in fact two global Gauss-Krüger zoned systems, one in six degree zones and one with three degree wide zones. The six degree version is very similar to UTM, having the same longitude of origin for each zone, but the zone numbering begins at the Greenwich meridian rather than 180°. Gauss-Krüger zone 1 has a longitude of origin of 3° East, zone 2 at 9°E, through to zone 60 with longitude of origin at 3° West. In both the six degree and three degree global Gauss-Krüger systems no scaling is applied to the projection; that is, the scale factor at the origin is unity (Table 3.3).

It may sometimes be inconvenient to have zone boundaries along meridians. For the State Plane coordinate system in the United States, the limits of area of use for zones are set by the boundaries of administrative divisions of the State. In these cases,

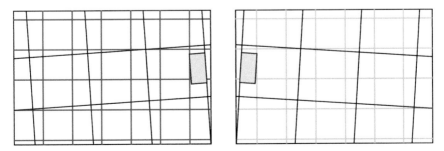

Figure 3.20 **A square feature on the surface of the Earth lying across the boundary between two Transverse Mercator zones, each half being plotted on a separate projection.**

Table 3.2 Projection parameters and their values for selected zoned TM projections.

Projection		Projection parameter values			
	Latitude of origin φ_0	Longitude of origin λ_0	Scale factor at origin k_0	False easting FE	False northing FN
Italy zone 1	0°N	9°W	0.9996	1 500 000.00 m	0.00 m
Italy zone 2	0°N	15°E	0.9996	2 520 000.00 m	0.00 m
Indiana State Plane CS83 West zone	37°30'N	87°05'W	0.999966667	900 000.00 m 2 952 750.000 ftUS	250 000.00 m 820 208.333 ftUS
Indiana State Plane CS83 East zone	37°30'N	85°40'W	0.999966667	100 000.00 m 328 083.333 ftUS	250 000.00 m 820 208.333 ftUS

the scale each side of a boundary will not necessarily be similar. In some States, such as Alabama, the 'zone' refers to area of coverage rather than commonality of map projection defining parameters (Table 3.4). In these cases, the zones are discrete Transverse Mercator projections.

The concept of zoning is one that has been introduced with reference to a particular projection method but, as with other projection concepts, is generally applicable for other methods.

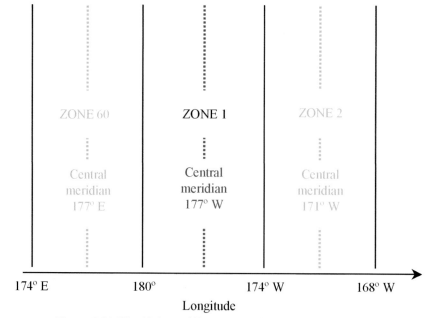

Figure 3.21 The Universal Transverse Mercator zone system.

Table 3.3 Projection parameters and their values for global zoned TM systems.

Projection	Projection parameter values				
	Latitude of origin φ_O	Longitude of origin λ_O	Scale factor at origin k_O	False easting FE	False northing FN
UTM ($\varphi > 0°$)	0°N	zonal	0.9996	500 000.00 m	0.00 m
UTM ($\varphi < 0°$)	0°N	zonal	0.9996	500 000.00 m	10 000 000.00 m
Gauss-Krüger System	0°N	zonal	1.0	500 000.00 m	0.00 m

3.3.5 South-oriented Transverse Mercator

A variation of the zoned Transverse Mercator is used in southern Africa in which the central meridian is orientated towards the *south* pole. In the Transverse Mercator method, eastings increment towards the east and northings towards the north. In the South-oriented method, the grid values increase towards the west and south. The formulae for the Transverse Mercator require minor modification to cope with this case (OGP 2007b). Conversely, the normal Transverse Mercator formulae cannot be used with the South-oriented Transverse Mercator parameter values. However, these projection parameters remain the same as for the normal Transverse Mercator, as shown in Table 3.5. In these examples, because the false easting values are defined as zero, negative grid coordinates are found to the east of the longitude of origin. (Negative coordinates north of the equator are outside the projection area of use.)

3.3.6 Oblique Mercator

The final classification of cylindrical is the situation where a country or region to be mapped is longer in one direction than another but is not aligned along a meridian or parallel. In this situation, it is possible to formulate an oblique aspect of the Mercator projection to minimise the scale factor, as in Figure 3.22.

Table 3.4 Projection parameters and their values for Alabama TM zones.

Projection	Projection parameter values				
	Latitude of origin φ_O	Longitude of origin λ_O	Scale factor at origin k_O	False easting FE	False northing FN
Alabama State Plane CS83 West zone	30°00'N	87°30'W	0.999933333	700 000.00 m	0.00 m
Alabama State Plane CS83 East zone	30°30'N	85°50'W	0.99996	200 000.00 m	0.00 m

Table 3.5 Projection parameters and their values for selected South-oriented TM projections.

Projection	Projection parameter values				
	Latitude of origin φ_O	Longitude of origin λ_O	Scale factor at origin k_O	False easting FE	False northing FN
South African Survey Grid zone Lo21	0°N	21°E	1.0	0.00 m	0.00 m
South West African Survey Grid zone Lo 22/21	22°S	21°E	1.0	0.00 GLM	0.00 GLM[1]

[1] German Legal Metre. I GLM = 1.0000135965 m.

In defining this projection method, it is necessary to specify the azimuth of the central line, as well as all the other parameters discussed before.

The scale factor will now be proportional to the secant of the angular distance from the centre line.

An example of the use of this projection method is in peninsular Malaysia, where the projection is termed the Hotine Oblique Mercator, or Rectified Skew Orthomorphic. The Alaskan panhandle also uses this projection method.

A rather similar projection method is the *Space Oblique Mercator,* but because this cannot be geometrically described it is dealt with separately in section 3.6.

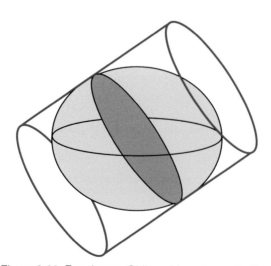

Figure 3.22 Forming an Oblique Mercator projection.

Table 3.6 Projection parameters and their values for selected
Oblique Mercator projections.

Parameter		Projection parameter values	
		Peninsular Malaysia RSO	Alaska State Plane CS83 zone 1
Latitude of projection centre	φ_C	4°N	57°N
Longitude of projection centre	λ_C	102°15'E	133°40'W
Azimuth of initial line	A_O	323°01'32.8763"	323°07'48.3685"
Angle from Rectified to Skew Grid	γ_C	323°07'48.3685"	323°07'48.3685"
Scale factor on initial line	k_O	0.99984	0.9999
False easting	FE	804 671.00 m	5 000 000.00 m
False northing	FN	0.00 m	−5 000 000.00 m

3.4 Azimuthal projections

3.4.1 General azimuthal

An azimuthal projection is formed by bringing a plane into contact with the sphere or ellipsoid and formulating a set of rules for the transfer of features from one surface to the other. Once again, the properties preserved can be distance, area, shape, or others.

Since the point of contact between a sphere and a plane is a single point, the scale factor distortion will be circularly symmetric. That is, the scale factor will be proportional to the distance from the centre of the projection. An azimuthal projection is therefore particularly suited to small 'circular' features on the surface of the Earth.

A special case of the azimuthal projection is where the point of contact is one of the poles. This is referred to as a *polar projection*. This has the rather obvious application of mapping the polar regions.

The polar projections are formed by taking the meridians off the sphere and placing them on the plane. The amount of distortion of the meridians will be a function of the type of projection, with the distortion of the parallels following in consequence. The general form of the polar projection will therefore be a set of meridians radiating from the pole with no distortion of the angle at the centre. This is shown in Figure 3.23.

As with the cylindrical projections, the azimuthal projections can have a further overall scaling applied to them, which has the effect of reducing the scale at the centre to less than 1, and making the scale true along a circle centred on the projection point. For the polar aspect this will make the scale true along a parallel of latitude away from the pole.

3.4.2 Azimuthal equidistant

The azimuthal equidistant is formed by keeping the scale factor equal to 1 in the direction radial from the centre of the projection. In the case of the polar equidistant, an example of which is shown in Figure 3.24, this means that the scale factor on the meridians, k_M, is equal to 1.

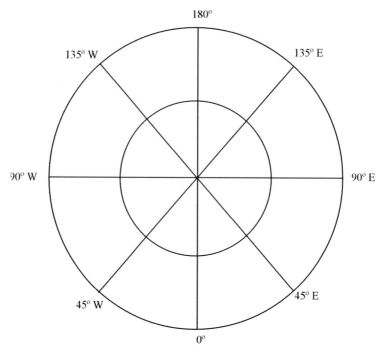

Figure 3.23 General form of the polar projection. Distances of parallels from the centre are a function of the projection method.

The scale factor along a parallel, k_p, is given as a function of latitude, ϕ, by:

$$k_P = \frac{\frac{\pi}{2} - \phi}{\cos \phi} \qquad (3.25)$$

and thus increases from 1 at the pole to 1.02 at 70° and 1.09 at 50°.

As a further example, Figure 3.25 shows an azimuthal equidistant projection that is centred on London. All distances from London are correct when measured from the map; all other distances will be too long.

3.4.3 Azimuthal equal area

The azimuthal equal area projection is formed in a similar way to the azimuthal equidistant, except that the scale factor of the lines radial from the centre is set to the inverse of the scale factor in the perpendicular direction.

For the polar aspect, shown in Figure 3.26, this leads to scale factors of:

$$k_M = \cos(45° - \phi/2) \qquad (3.26)$$

$$k_P = \sec(45° - \phi/2) \qquad (3.27)$$

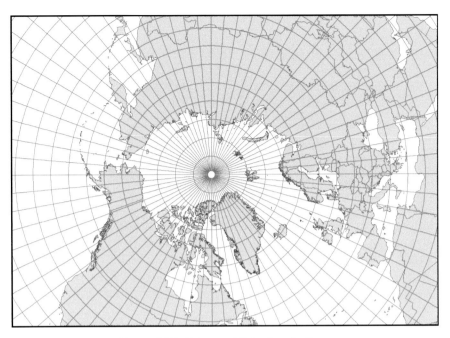

Figure 3.24 **Polar equidistant projection.**

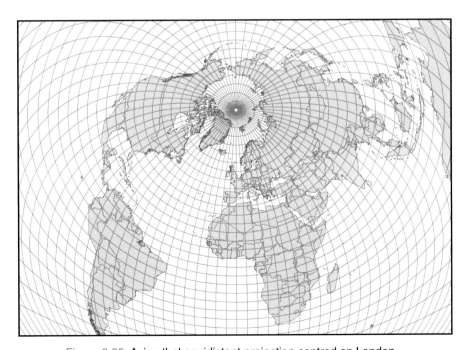

Figure 3.25 **Azimuthal equidistant projection centred on London.**

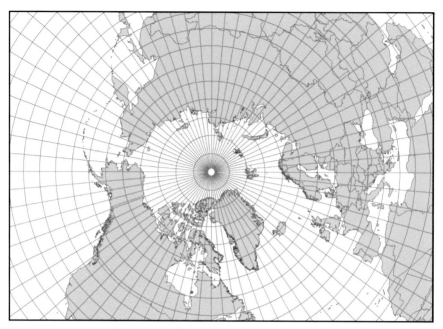

Figure 3.26 The polar equal area projection.

The ellipsoidal formulae are more complex than equations 3.26 and 3.27. A projection based on the ellipsoidal form developed by Lambert is used for European statistical mapping and other purposes where true area representation is required. Parameter values are given in Table 3.7 (Annoni *et al.* 2001).

3.4.4 Stereographic

The conformal version of the azimuthal projection is termed the *stereographic.* This is for historical reasons, since this projection can be constructed graphically by projecting all points from a 'viewing point' on the opposite side of the Earth from the centre of the projection as shown in Figure 3.27.

As with all conformal projections, this one has a particular significance as it is sometimes used as the basis for national mapping, being particularly appropriate for small, compact countries or islands. The polar stereographic, shown in Figure 3.28, is used as a complement to the Universal Transverse Mercator system beyond latitudes

Table 3.7 Projection parameters and their values for Europe Lambert Equal Area projection.

Projection	Projection parameter values			
	Latitude of origin φ_o	Longitude of origin λ_o	False easting FE	False northing FN
Europe Lambert Equal Area 2001	52°N	10°E	4 321 000.00 m	3 210 000.00 m

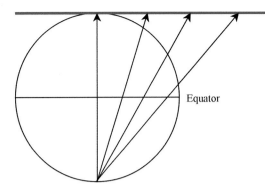

Figure 3.27 **Formation of a stereographic projection.**

84°N and 80°S, when it is known as the Universal Polar Stereographic projection, or UPS. In this usage, the scale at the pole (k_O) is reduced to 0.994, which results in a *standard parallel* (where scale is true) of 81°06'52.3". Other parameter values are given in Table 3.8. This *secant* case of the formation of a stereographic projection is illustrated in Figure 3.29.

In general, the scale factor for the polar aspect is given by:

$$k = k_O \sec^2(45° - \varphi/2) \qquad (3.28)$$

which is the same in any direction.

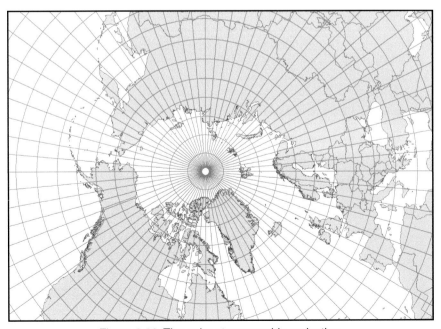

Figure 3.28 **The polar stereographic projection.**

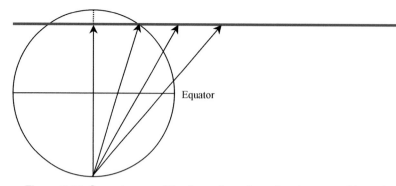

Figure 3.29 Secant case of the formation of a polar stereographic projection, with the scale true at a parallel of latitude and not the pole.

The oblique aspect of the stereographic projection is tangential at a chosen point on the surface of the sphere or ellipsoid. If a scale factor is applied, the scale is true on (in the spherical case) a small circle centred on the origin. The projection method is used in its ellipsoidal form for the mapping of some countries and provinces where the area covered has similar ranges of latitude and longitude. Examples are given in Table 3.9.

For the ellipsoidal development, the ellipsoidal coordinates are first transformed into spherical values and then projected onto the grid. Note that there are two approaches to the first step. One converts the ellipsoidal coordinates onto a conformal sphere with a radius based on the ellipsoidal radii at projection origin, whilst the alternative uses a radius based on the ellipsoidal radii at each point. Away from the origin the two methods give diverging results, differing by about 1 m at 100 km. It is therefore essential to identify the correct method and use the appropriate formula. Unfortunately both methods tend to be called Oblique Stereographic and some local knowledge is needed to identify that required for local use. The generally excellent formulae given in Snyder (1987) take the former approach, whilst most practical cases outside of the US, including all of those in Table 3.9, take the latter. This latter

Table 3.8 Projection parameters and their values for the Universal Polar Stereographic projection.

Projection	Projection parameter values				
	Latitude of origin φ_o	Longitude of origin λ_o	Scale factor at origin k_o	False easting FE	False northing FN
Universal Polar Stereographic North	90°N	0°E	0.994	2 000 000.00 m	2 000 000.00 m
Universal Polar Stereographic South	90°S	0°E	0.994	2 000 000.00 m	2 000 000.00 m

Table 3.9 Projection parameters and their values for selected Oblique Stereographic projections.

Projection	Projection parameter values				
	Latitude of origin φ_O	Longitude of origin λ_O	Scale factor at origin k_O	False easting FE	False northing FN
Netherlands RD	52°09'22.178"E	5°23'15.5"E	0.9999079	155 000.00 m	463 000.00 m
Romania Stereo 70	46°N	25°E	0.99975	500 000.00 m	500 000.00 m
Prince Edward Island Stereographic	47°15'N	63°00'W	0.999912	400 000.00 m	800 000.00 m

method is sometimes called the 'Double Stereographic' in North American coordinate conversion applications. In fact, several applications that originate from the US Geological Survey (such as the PROJ.4 conversion software) use Snyder's formula, and this problem of the different versions of the ellipsoidal stereographic projection has been inherited by them. We repeat here the warning that we gave in the introductory sections of this chapter: *use the models, formulae, and parameters that have been specified for the coordinate reference system in which you are working*, or there may be significant errors.

3.4.5 Gnomonic

One particular form of the azimuthal projections should be noted here, as although it is seldom used in modern applications, it does have some interesting properties. This is the gnomonic projection, which is formed by projecting all points from the centre of the Earth as shown in Figure 3.30.

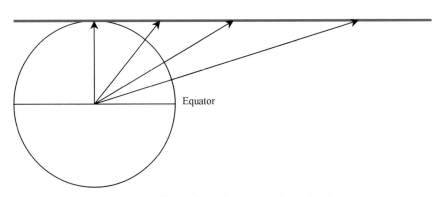

Equator

Figure 3.30 Formation of a gnomonic projection.

An example of the polar aspect is shown in Figure 3.31. As would be expected, the scale factor distortion becomes extreme away from the centre of the projection, reaching a value of 2 along the meridian when the latitude is 45° and a value of 4 at a latitude of 30°. It is clearly not possible to show an entire hemisphere with this projection.

The only advantage of this projection is that it is the only one where all great circles (the shortest route between two points on a sphere) are shown as straight lines on the projection, and vice versa. This feature means that it can be used to plan the shortest route between two points, although this role has largely been superseded by computational techniques.

3.4.6 Azimuthal orthographic

The azimuthal orthographic projection is formed by projecting all points on the ellipsoid onto a plane, along lines that are normal to the plane. The technique is illustrated in Figure 3.32, and an example of a resulting projection is shown in Figure 3.33.

The resulting depiction of the Earth is analogous to how it would appear from deep space using a powerful telescope. It is mainly of interest as a special case of the perspective projection covered in the next section.

3.4.7 Azimuthal perspective projection

The azimuthal perspective projection is similar to the azimuthal orthographic, except that it represents how the Earth would be seen from a finite view point. Its formation is illustrated in Figure 3.34.

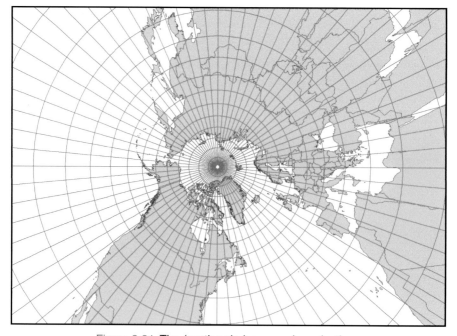

Figure 3.31 The (north polar) gnomonic projection.

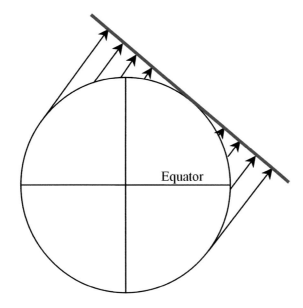

Figure 3.32 Formation of an azimuthal orthographic projection.

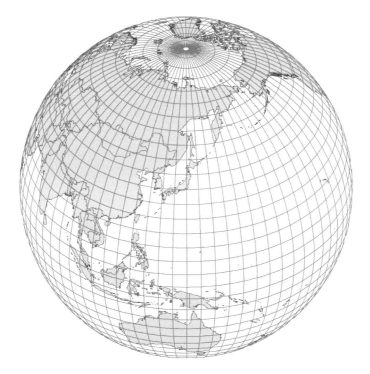

Figure 3.33 **Example of an azimuthal orthographic projection, here centred on Tokyo.**

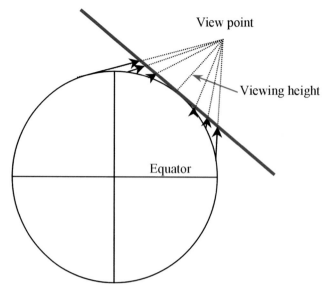

Figure 3.34 **Formation of an azimuthal perspective projection.**

The orthographic version is a special case in which the viewing height is infinity. For the general case, less than a hemisphere can be presented on any single view.

This projection is of interest as it is the one used by some web-based mapping systems such as Google Earth™. In connection with this, several points can be noted:

- The scale is true at the centre of the projection. Since the projection is not conformal, the scale factor will be different along the radial and circumferential directions. The formulae for the spherical model are:

$$k_{RADIAL} = \frac{h_V}{h_V + R(1 - \cos\theta)} \frac{\sin\theta}{\theta} \tag{3.29}$$

$$k_{CIRCUM} = \frac{h_V}{h_V + R(1 - \cos\theta)} \tag{3.30}$$

- In these formulae, θ is the angular distance away from the centre in radians, R is the radius of the Earth, and h_V is the viewing height. R and h_V need to be in the same units.
- Distances inferred from coordinates in such a projection would be in error according to the value of the scale factor. However, Google Earth™ does not use projections when computing the distance between points. Instead, it takes the points identified on screen and converts them back to geographic coordinates, and then computes the great circle distance between them using spherical trigonometry. The formulae for this type of computation are given in Appendix C.

- In fact, the projection used by Google Earth™ is not a pure perspective azimuthal, since it takes account of the height of the terrain. It will be inferred from Figure 3.34 that a point above the surface of the ellipsoid has its position misplaced along the radial direction compared to a point on the ellipsoid. The amount of displacement will depend upon the viewing height, the terrain height, and the position of the point relative to the centre. As the viewing height increases, the effect of terrain is less significant, and so when viewing entire continents there is no noticeable difference to the case without terrain heights. However, as the view height decreases, the terrain can become significant, and when panning across mountains at low altitudes the user will notice the summits moving across the screen at a different rate to the valleys. There is no distortion due to terrain height at the point directly in the centre of the screen.

    At these low viewing altitudes, the proportional displacement from the centre (as a proportion of the un-distorted height) is given approximately by the ratio $(h_V + h_T)/ h_V$, where h_T is the height of the terrain in the same units as the viewing height h_V.

- If attempting to produce an overlay in Google Earth™ using another data set, it is tempting to think that the geometry of the azimuthal perspective projection will have an effect. However, although this is the projection used to view and manage the process, it is the geometry of the other projections that actually comes into play. Section 4.5.7 covers this type of operation.

3.5 Conic projections

3.5.1 General conic

A conic projection is formed by bringing a cone into contact with the sphere or the ellipsoid. In so doing, it is seen to be touching the sphere along a parallel of latitude. This line is known as the *standard parallel* of the projection.

It can be seen in Figure 3.35 that many different shapes of cone can be selected, all resulting in a different standard parallel. The choice will depend upon which region

Figure 3.35 **Cones in contact with different standard parallels.**

of the Earth is to be mapped, an appropriate standard parallel being one that passes through the centre of the region.

The resultant form of the conic projection is that the meridians appear as straight lines converging towards one of the poles. The angle between two meridians is a function of the standard parallel, and can be expressed as:

$$\gamma = \Delta\lambda\sin\alpha \tag{3.31}$$

where $\Delta\lambda$ is the difference in longitude of the two meridians and α is the latitude of the standard parallel.

The conic is in fact the general case of projection of which the cylindrical and azimuthal projections are limiting forms. In equation 3.31, as α tends towards 90°, γ tends to $\Delta\lambda$, which indicates that the angles between the meridians are true. This is as in the polar projection, which is the equivalent of a completely flat cone touching the sphere at the pole.

Similarly, as α tends to 0°, γ tends to 0°, which indicates that the meridians are parallel as in a normal conic projection. Again, a cylinder is the equivalent of a cone touching the equator.

These considerations are useful for gaining an insight into the nature of conic projections, but should not be implemented in practice, as the formulae for the cone are very likely to break down under these extreme conditions.

The equivalent of an overall scaling is often used for conic projections, where it is achieved by using *two standard parallels* as shown in Figure 3.36. The effect is to reduce the scale factor below 1 in between the two standard parallels and increase it above 1 outside them. It should be noted that for the ellipsoidal case of the projection method the minimum scale factor is not exactly at the parallel midway between the two standard parallels as for a sphere, but is slightly poleward of the mean.

Finally, it should be noted that for any conic projection the scale factor is entirely a function of latitude, and these projections are therefore suitable for depicting regions with a broad extent in longitude, particularly those regions in mid-latitudes.

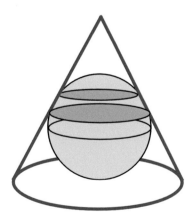

Figure 3.36 Formation of conic projection with two standard parallels.

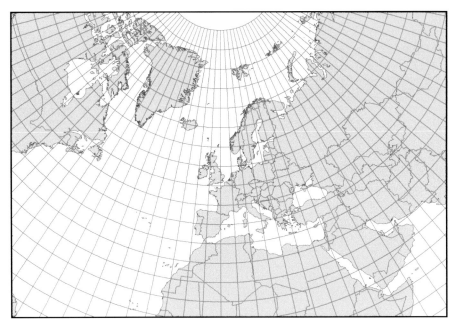

Figure 3.37 Conic equidistant projection, standard parallels at 20° and 60° North.

3.5.2 Conic equidistant

A conic equidistant projection preserves the scale factor along a meridian ($k_M = 1$). The parallels are then equally spaced arcs of concentric circles. The scale factor along a parallel of latitude is given as a function of latitude φ by (Snyder 1987):

$$k_p = \frac{(G - \varphi)n}{\cos \varphi} \tag{3.32}$$

where

$$G = \frac{\cos\varphi_1}{n} + \varphi_1 \tag{3.33}$$

$$n = \frac{\cos\varphi_1 - \cos\varphi_2}{\varphi_2 - \varphi_1} \tag{3.34}$$

and φ_1 and φ_2 are the two standard parallels. If only one standard parallel is used, then n in equation 3.34 is equal to $\sin\varphi_1$.

An example of this projection is shown as Figure 3.37.

3.5.3 Albers equal area

The equal area version of a conic projection is usually called Albers equal area. An example of this for the European region is shown in Figure 3.38.

It will be noted that the pole is shown in this projection as a circular arc, indicating once again that shape has been sacrificed to keep the area undistorted. It should also be noted, however, that the shape is not as badly distorted as in the cylindrical equal area projection in Figure 3.11. This is mainly a function of the region being projected: the European area shown on these two maps is by and large an area in

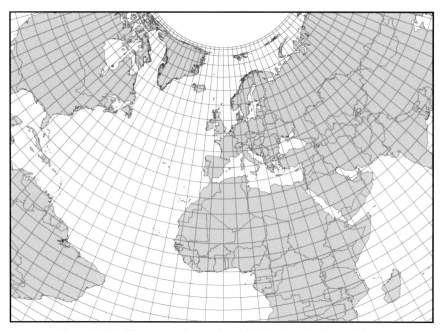

Figure 3.38 **Albers equal area (conic) with standard parallels at 20° and 60° N.**

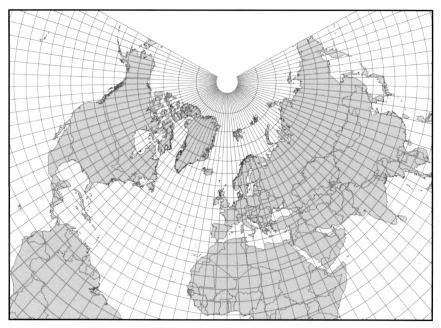

Figure 3.39 **Lambert Conformal Conic with standard parallels at 20° and 60° N.**

mid-latitude with a large east-west extent, which is more suited to a conic projection than a cylinder.

3.5.4 Lambert Conformal Conic

The conformal version of the conic projection is usually named after Lambert, who first developed it in 1772 (Snyder 1987). The full name is the Lambert Conformal Conic (LCC) although most references to a Lambert projection would usually be understood to refer to this one (though not without some ambiguity as Lambert was responsible for developing several other projection methods such as the equal area projection described in section 3.4.3).

This is an extremely widely used projection, and it is probably true to say that the LCC and the Transverse Mercator methods between them account for 90% of base map projections world wide.

An example of the LCC is shown in Figure 3.39. Since it is a conformal projection, the meridians meet at a point, which represents the pole. The example in Figure 3.39 was formed with the equivalent of a standard parallel at 40°. A LCC with a standard parallel at the equator would effectively be the same as a Mercator projection, with the meridians parallel and never reaching the infinite pole; one with a standard parallel at 90° would be the same as a polar stereographic projection. Examples near these extremes are shown in Figure 3.40, but as remarked in the introductory section on conic projections, the LCC formulae should not actually be used with these extreme parameter values.

Parameters and their values for selected LCC projections are given in Table 3.10. Note that the three French projections form a zoned system, similar in principle to the Transverse Mercator zoning discussed earlier in section 3.3.4 in section 3.3.4. For the LCC,

Table 3.10 Projection parameters and their values for selected Lambert Conformal Conic with one standard parallel projections.

Projection	Projection parameter values				
	Latitude of origin φ_O	Longitude of origin λ_O	Scale factor at origin k_O	False easting FE	False northing FN
Jamaica Metric Grid 2001	18°N	77°W	1	750 000.00 m	650 000.00 m
France Lambert zone I	55 grads	0 grads E of Paris	0.999877341	600 000.00 m	1 200 000.00 m
France Lambert zone II	55 grads	0 grads E of Paris	0.999877420	600 000.00 m	2 200 000.00 m
France Lambert zone III	49 grads	0 grads E of Paris	0.999877499	600 000.00 m	3 200 000.00 m

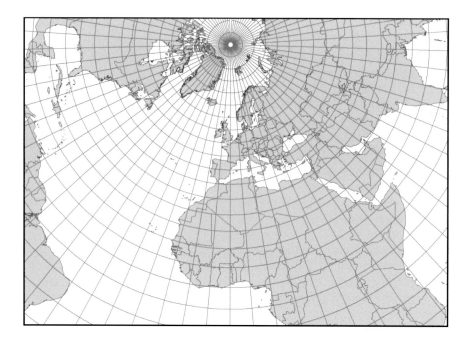

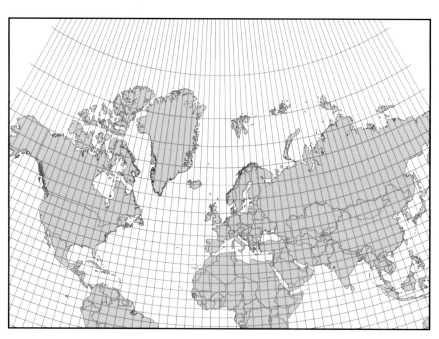

Figure 3.40 LCC projections with standard parallel at
80° (above) and 10° (below).

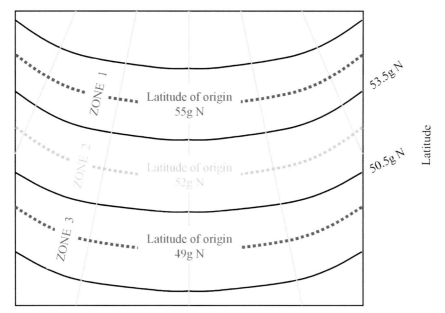

Figure 3.41 **A Lambert Conic Conformal zone system.**

the zones are tiered from north to south, as shown in Figure 3.41. The zone boundary is midway between the parallels of origin for adjacent zones. But unlike zoning on the Transverse Mercator, in order to make the scale factor for each zone equal on the zone boundary, the scale factor on the parallel of origin has to differ for the LCC.

An LCC projection may also be formed with two standard parallels, as with all conic projections. In this case, it is the equivalent of one standard parallel half way in between, with an additional scaling applied. The usual arrangement for minimising distortion would be to have two standard parallels which are each 1/6 of the range of latitude in from the extremes of the projection.

The formulae for LCC are complicated, but the expression for scale factor for the case with one standard parallel and a spherical Earth can be quoted as (Snyder 1987):

$$k = \frac{\cos\varphi_1 \, \tan^n\left(\frac{\pi}{4} + \frac{\varphi_1}{2}\right)}{\cos\varphi \, \tan^n\left(\frac{\pi}{4} + \frac{\varphi}{2}\right)} \tag{3.35}$$

where φ_1 is the standard parallel

and $n = \sin\varphi_1$ $\tag{3.36}$

Some examples of LCC projections with two standard parallels are given in Table 3.11.

3.5.5 Oblique conic
As with the cylinder, the axis of the cone does not necessarily have to be coincident with the axis of the Earth. Occasionally, an oblique aspect may be encountered, as shown in Figure 3.42.

Table 3.11 Projection parameters and their values for selected Lambert Conformal Conic with two standard parallels projections.

Projection	Projection parameter values					
	Latitude of origin φ_O	Longitude of origin λ_O	Latitude of first standard parallel φ_1	Latitude of second standard parallel φ_2	False easting FE	False northing FN
Europe Conformal 2001	52°N	10°E	65°N	35°N	4 000 000.00 m	2 800 000.00 m
Tennessee State Plane CS83	34°20'N	86°00'W	36°25'N	35°15'N	600 000.00 m 1 968 500.00 ftUS	0.00 m 0.00 ftUS
Northwest Territories Lambert	0°N	112°W	70°N	62°N	0.00 m	0.00 m
Geoscience Australia Lambert	0°N	134°E	18°S	36°S	0.00 m	0.00 m

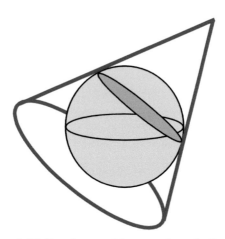

Figure 3.42 Forming an oblique conic projection.

Once again rules can be proposed to produce equal area, equidistant, or conformal projections as required. An ellipsoidal development of the conformal case is the Krovak projection used in the Czech Republic and Slovakia.

3.6 Non-geometric projection methods

The projection methods discussed so far have been characterised by being constructed on a developable surface – a cylinder, a plane, or a cone. With the advent of modern computing power and using more complex mathematics, it is now feasible to create projections in which the central line follows a feature of interest yet the desired equal area or conformal projection property is retained. In fact, we have already alluded to this in equation 3.2 of section 3.2.3, where we first introduced developable surfaces. That equation simply stated that grid coordinates can be written directly as a mathematical function of geographic coordinates. Of course, how complicated that function is depends to a great extent on how much computing power is available. In this section, we introduce some of the more significant examples of the non-geometrical projections of this kind, taking them in the order in which they were developed.

The New Zealand Map Grid was developed in the early 1970s (Reilly, 1973). The problem with the shape of New Zealand is that it does not fit neatly into any of the categories that we have identified in this chapter as indicating some natural choice of projection method: it is not predominantly north-south, nor spread east-west along a particular latitude, nor long and thin and oriented obliquely to the meridians. Of course it is not the only country on the Earth that cannot be neatly classified in this way, but the New Zealand Department of Lands and Surveys at the time decided that it was a problem that could be overcome. The solution adopted was closest in concept to an Oblique Mercator projection but, rather than being developed with respect to a straight line at a given azimuth, the lines of constant scale factor were complex curves that followed the approximate trend of the main islands of New Zealand.

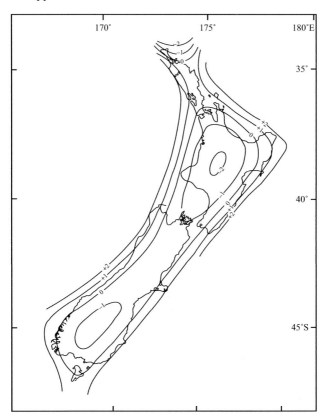

Figure 3.43 Scale factor distortion on the New Zealand Map Grid, from Reilly (1973). Contours are parts per ten thousand and range from +2 to −2.

The root mean square deviation of the scale factor from unity on the resulting grid was 120 ppm, with a range of ±200 ppm (Reilly, 1973).

Unfortunately, the downside of a development of this kind was that only mapping software and GIS packages that supported the use of this projection could be used, and with it only being applicable to New Zealand it was difficult to persuade software developers to add it to their functionality. Mapping authorities in New Zealand have therefore reverted to a conventional system based on the Transverse Mercator projection (LINZ 2001).

A later development was the Space Oblique Mercator projection developed by (Snyder 1981, 1987) to handle satellite imagery. A remote sensing satellite has a ground track that crosses the equator at a certain azimuth that gradually increases to 90° when the ground track reaches its maximum latitude. Thereafter the ground track heads south with increasing azimuth, before crossing the equator and repeating this pattern in the southern hemisphere. From the point of view of the satellite images, the line along which scale is true is therefore a complex curve that oscillates between northern and southern extremes of latitude and goes through many such cycles before repeating itself. Snyder therefore developed the Space Oblique Mercator to describe this property and allow a relationship to be made to geographic coordinates. The parameters of the projection include the orbital elements of the satellite's ephemeris (inclination, orbital period, right ascension of the ascending node, and so on) but *locally* it appears similar to an ordinary Oblique Mercator. We could therefore use the insight from section 3.3.6 to draw such conclusions as the scale factor being approximately equal to the secant of the angular distance from the centre line.

The most recent development of those discussed here was the Snake projection developed by (Iliffe *et al.* 2007) for use on railways and pipelines. It will have been apparent from equation 3.1 (and it is further discussed in section 3.8) that there is a difference between a distance measured on the surface of the Earth and one implied by the coordinates of a projected coordinate reference system. For many engineering applications, the standard procedure is for the surveyor or engineer to make scale factor and reduction to ellipsoid corrections before transferring a measured distance onto the projection. However, in some industries, and especially in modern infrastructure work on railways with its use of automated track guidance machines and emphasis on the integrity of coordinate geometry, there is a reluctance to adopt this approach.

For some engineering projects it is possible to tailor a conventional projection so that the local scale factor is close to unity across the site – for example using a Transverse Mercator with a natural origin in the centre of the site. However, for a railway that extends several hundred kilometres, this is generally not an option. The Snake projection therefore developed the concept of making the scale factor unity along a gently sinuous line (hence the name) that followed the trend of the railway (or other largely linear construction project). The term *gently* sinuous is used to indicate that sharp changes of direction of the trend line are avoided in order not to distort the conformal properties of the projection; in fact, since there is no need to follow the exact route of the railway, a trend line can usually be found with a radius of curvature above 100 km or so that keeps to within 10 km of the actual track. A further feature of the Snake projection is that it also uses a *vertical* trend line to keep to the same approximate height of the railway, and thus does away with the need for corrections to sea level or ellipsoid as well as eliminating the need for scale factor corrections.

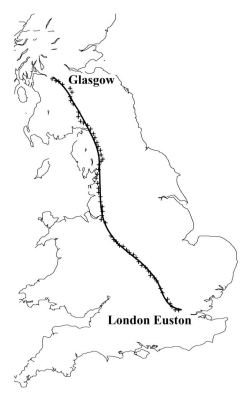

Figure 3.44 **The Snake projection as applied to the West Coast Main Line from London to Glasgow. The crosses show the generalised route of the track and the continuous line is the trend line with unity scale factor. (After Iliffe *et al.* 2007)**

Using this projection in its first application, a trend line was fitted to the West Coast Main Line from London to Glasgow to within 9 km and with a minimum radius of curvature of 40 km. This is shown in Figure 3.44, along with the actual path of the railway. It was found that the maximum deviation from unity of the scale factor on the actual track was 11 parts per million, along a route that stretched for 645 km and passed over hills up to 300 m above sea level.

3.7 Summary of information required

3.7.1 Map projection method formulae

Examples of formulae for converting between geographic coordinates and projected coordinates were given in section 3.3.1 for the cylindrical equidistant projection. Throughout this book, however, the assumption has been made that the vast majority of users will not be called on to program specific algorithms into a computer, but will have access to a wide variety of projection methods via a suite of software. Some applications support over forty different projection methods. Some of these may be individual projected coordinate reference systems and include the datum, projection method and projection parameter values in the specification; others may be generic projection methods that will require parameter values to be input.

It may occasionally prove useful to have access to a set of programs for projection computations that are independent of a particular application. One such is the PROJ.4 package from the USGS (United States Geological Survey), which is available over the World Wide Web. The location is given in the References under 'PROJ.4'.

If it is necessary to use a particular projection, for which no computer program is available, then it will be necessary to consult a standard work of reference. One of the best available is Snyder (1987), which has already been referred to in some of the formulae given in preceding sections. This work contains formulae in spherical and ellipsoidal form for over 30 main projection methods, and includes a comprehensive set of worked examples for each. For certain methods, OGP (2007b) has developed Snyder's work further for more general global application.

3.7.2 Map projection parameter values

Each projection method is defined both in terms of the formulae needed for converting between geographic and grid coordinates, and the parameters of the method that are necessary as input for those formulae.

Except in the case of standard projections such as UTM and UPS, a projection is not defined simply by its type. A Transverse Mercator projection with central meridian 2°W, for example, is completely different to a Transverse Mercator with central meridian 20°E.

Does it matter what the parameters are and what projection has been used? A map may be digitised in grid or projected coordinates and used in much the same way that a paper map would be used. However, as soon as it is necessary to carry out any kind of computation based on the information presented in the map, then a knowledge of the projection, its method, and its parameter values is necessary. This may be a simple computation such as the area of a feature on the map or the distance between two points, in which case the projection may be ignored for low accuracy applications. But if it is required to combine the data from the map with other data, perhaps from a satellite image or from additional information obtained and referenced using GPS, then it is necessary to transform all information into a common coordinate reference system. For this, a knowledge of the projection and its parameters is required.

Table 3.12 summarises the information that is usually necessary for each projection method. The meaning of these terms is summarised in Table 3.13.

It will be noted that the conversion formulae given earlier in this chapter include the radius of the sphere. Ellipsoidal formulae require parameters describing the size and shape of the ellipsoid. The sphere or ellipsoid parameters are part of the geodetic datum definition for the geographic coordinate reference system to which the projection is applied. This is discussed in Chapter 2.

The question then remains of where to obtain this projection and datum information. Some maps have the name of the projection written in the margin, either specifically or in rather a vague way (eg: 'conic with two standard parallels'). It is very unusual, however, to see the actual values of the parameters printed on a map.

Most national mapping organisations publish the values of the parameters that are used for their own map series, for example (Ordnance Survey 2006), but obtaining the information may be a lengthy process. Nor does it mean that all maps of that country will be printed in that projection: a publisher may have devised their own, and it is also difficult to obtain this information.

Table 3.12 Parameters necessary to define each projection.

Projection method	Parameters required by method								
	φ_O	λ_O	φ_1	φ_2	A_O	γ_C	k_O	FE	FN
Cylindrical equidistant	✓	✓					✓	✓	✓
Cylindrical equal area	✓	✓					✓	✓	✓
Mercator	✓	✓					✓	✓	✓
Transverse Mercator	✓	✓					✓	✓	✓
Oblique Mercator	✓	✓			✓	✓	✓	✓	✓
Azimuthal equidistant	✓	✓					✓	✓	✓
Azimuthal equal area	✓	✓						✓	✓
Stereographic	✓	✓					✓	✓	✓
Gnomonic	✓	✓					✓	✓	✓
Conic equidistant	✓	✓	✓	✓				✓	✓
Albers equal area	✓	✓	✓	✓				✓	✓
Lambert Conic Conformal (1SP)	✓	✓					✓	✓	✓
Lambert Conic Conformal (2SP)	✓	✓	✓	✓				✓	✓

Table 3.13 Summary of meaning of terms in Table 3.12.

φ_O	Latitude of the true origin. Not necessarily the same as φ_1 for the conic projections.
λ_O	Longitude of the true origin. Equivalent to the central meridian for cylindrical and other projections.
φ_1	The first (or only) standard parallel.
φ_2	The second standard parallel.
A_O	The azimuth of the centre line for oblique projections.
γ_C	The rectifying rotation of the map grid for oblique projections.
k_O	Overall scaling factor to be applied. May be referred to as the central meridian scale factor for Transverse Mercator. Not usually applied to conic projections, as its role is performed by using two standard parallels.
FE	False eastings to be added to all coordinates. Equivalent to the eastings at the true origin. May alternatively be referred to as E_O or E_F.
FN	False northings to be added to all coordinates. Equivalent to the northings at the true origin. May alternatively be referred to as N_O or N_F.

A reference such as the EPSG geodetic parameter dataset (OGP 2007a) may prove useful. Hints on using such repositories are given in section 2.5.2.

3.8 Computations within map projections

Aside from the conversions between projected coordinates and geographic ones, or from one projected coordinate system to another, we can identify two principal types of computations that will generally need to be carried out in association with map projections:

- Calculating real-world properties such as distance, area, and bearing from coordinates that are given in a projected form, and have thus been subjected to a degree of distortion;
- In the opposite sense, transferring survey measurements of distances and angles on the Earth to their projected equivalents, so that surveys can be computed in the simpler two-dimensional geometry of a projection rather than having to use geographic coordinates.

In considering the amount of distortion between real-world properties and their equivalents on the projection, we can re-cast equation 3.1 of section 3.2.2 into either of the two forms of equations 3.37 and 3.38.

$$\text{distance on the projection} = (\text{distance on the ellipsoid}) \times k \qquad (3.37)$$

$$\text{distance on the ellipsoid} = (\text{distance on the projection}) / k \qquad (3.38)$$

For any application, the key is therefore the scale factor, k, and this needs to be related to the degree of error that can be tolerated. For a map projection that has been designed to represent the whole Earth, we have seen in the preceding sections that k can be very large indeed – values up to infinity are encountered somewhere on most projections. So although we might identify certain exceptional cases where the effect of scale factor is limited in particular circumstances, we can generally say that computations made on projections designed for the whole Earth are fraught with difficulty and are to be avoided. The exceptional cases are such as:

- Distances computed from the central point on an azimuthal equidistant projection are undistorted. For example, such a projection might be used to show airline routes from a central hub such as London Heathrow, and all distances from this point can be measured correctly. The distance from Cape Town to New York would be completely wrong, however.
- Areas computed on an equal-area projection are correct. However, it should be remembered that distances are then wrong.

The more common case will be that we are dealing with map projections that have been designed to display topographic mapping on paper or in digital form. It is generally the case that the areas that these cover have been restricted in order to keep the distortion of scale factor away from unity within reasonable limits: usually within around ± 0.1%.

With this type of distortion we can immediately see that for some applications we need not worry at all, and can simply use projected coordinates as if they truly represent the Earth as it is. Examples might be computing the distance of an emergency response vehicle from the scene of an accident in order to estimate its time of arrival, or computing the population that lives within a given radius from a supermarket. In each of these cases, there are far more signi⬚cant problems to consider, even in the unlikely event that we really wanted an answer to an accuracy better than 0.1%. In fact, such applications could probably tolerate the 2–3% distortion that would result from representing a country the size of the USA or Australia on a single projection.

Another example would be computing the area of a farmer's ⬚eld, perhaps for land valuation or to estimate the crop yield. In determining areas we should note that the scale factor applies in two dimensions, and therefore the previous equation 3.38 needs to be adapted to:

$$\text{area on the ellipsoid} = (\text{area on the projection}) / k^2 \qquad (3.39)$$

Note that in equation 3.39 the assumption is that we are dealing with a conformal projection, which will nearly always be the case when dealing with medium or large scale topographic mapping, and hence the scale factor at any one point will be the same in all directions. For an equal area projection, the scale factor in one direction will be the reciprocal of the scale factor in the perpendicular direction, and equation 3.39 would reduce to the equal area condition.

For the farmer trying to compute the amount of fertiliser that will be required in each ⬚eld, the error would therefore be up to 0.2% if using a typical national coordinate reference system, which is unlikely to be a serious problem. On the other hand, there are situations when this would not be acceptable. In the oil industry, for example, some licence areas cross national boundaries and a process of 'unitisation' needs to be carried out to determine the proportion owned by each state – with large sums of money involved this needs to be carried out accurately. The options are either to transform all the data into an equal-area projection or to compute the scale factor that is applicable to each part.

A rather more stringent accuracy requirement is usually associated with traditional land surveying, where we use equation 3.37 to transfer distances measured on the Earth to the map projection. In this context, 0.1% translates as a distance of 100 m having to be adjusted by 10 cm in order to match the projected coordinates, which is certainly signi⬚cant when compared with the accuracy that may be obtained with an electromagnetic distance measurer (EDM). We should also note here that equation 3.37 applies strictly to the ellipsoid, and not the actual location where the distance was observed. Therefore there is another step to be applied, in order to project a distance observed at a height h down (or up if the ellipsoidal height is negative) to its equivalent on the ellipsoid:

$$D_{ELLIPSOID} = \frac{R}{R+h} D_{MEASURED} \approx D_{MEASURED} - \left(\frac{h}{R}\right) D_{MEASURED} \qquad (3.40)$$

In this equation, R is a suitable value for the radius of curvature of the ellipsoid: a spherical assumption can often be made, since the correction is less than 2 cm per hundred metres for heights up to 1000 m.

For the scale factor correction, it was pointed out in section 3.2.2 that the de⬚nition of scale factor applies in theory to lines of in⬚nitesimally short length. In practice, the scale factor changes so slowly across a projection that a single scale factor can often be considered as applicable to all the distances in one survey, rather than having to compute a separate one for each. How slowly? In part, the answer will depend on the particular projection, but as the area that any one survey projection covers is restricted in order to avoid excessive corrections of this type, a Transverse Mercator projection applicable to a 6° wide zone may be taken as a suitable example and inferred as being typical of most survey projections.

The region with the most extreme rate of change of scale factor is on the edge of the projection. Over a distance of 5 km, the scale factor may, for example, vary from 1.00043 at one end of a line to 1.00039 at the other. The error introduced by using the scale factor at one end of the line, rather than the average computed over its whole length would in this situation be around 8 cm. A similar calculation over a distance of 3 km in the worst case scenario shows a potential error of 3 cm.

For a target accuracy of 1 cm, and distances over 1 km, it would therefore be advisable to calculate a scale factor for each line. This can either be done by using the mean of the end point values of scale factor, or by using Simpson's rule (Allan 1997a) for higher precision:

$$k_{AB} = \frac{1}{6}(k_A + 4k_M + k_B) \tag{3.41}$$

In equation 3.41, k_{AB} is the line scale factor, k_A and k_B are the end point values, and k_M is the scale factor for the mid point. For a project spread over an area of less than 1 km² , it will nearly always be acceptable for a general scale factor to be used for the site as a whole.

Survey measurements typically involve observations of angles and distances, and in fact the former are distorted on a projection as well as the latter. The correction to be applied to angles arises because, in general, the straight line observed between two points does not plot as a straight line on the projection. Standard trigonometric calculations on the coordinates, distances, and angles assume that all lines are straight, and so a correction has to be applied to obtain the angle that *would* have been observed along the straight lines on the projection. This is illustrated in Figure 3.45.

The magnitude of this correction is proportional to the length of the line, and is never large. At its most extreme on the edge of a Transverse Mercator of zone width ±3° from the central meridian it reaches 3 s over a distance of 5 km, and 6 s over 10 km. Its significance has greatly diminished with the use of GPS in place of triangulation over long distances. For the vast majority of modern surveys, it may therefore be safely dismissed. Otherwise, a text such as (Allan 1997b; 2007) may be consulted.

Finally, for the sake of completeness in describing the differences between features on a projection and on the real Earth, we remind readers of the property of *convergence* that was discussed in section 3.3.4. This is the angle between grid north and true

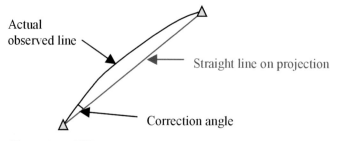

Figure 3.45 Difference between actual line and projection line.

north, and even on the type of survey projections that we have been discussing in the latter part of this section can reach 2–3°.

Case Study 6.4 gives an example of dealing with problems that arise when carrying out computations using projected map coordinates.

3.9 Designing a map projection

For many applications, those working with map projections will be using an existing projected coordinate reference system, and the task will be to identify the projection method and the associated parameters. A different kind of situation is one where it is necessary to design a projection for a particular purpose, in which case the choice of projection method and parameters are for the designer to decide.

In such a case, it is first necessary to define what is meant by a suitable projection. A suggested order of criteria is given below.

1. Firstly, it should preserve any properties that the use of the map dictates. That is, if the map or coordinate reference system is to be used to measure or compute the areas of features, then it may help if the projection were an equal area projection. Alternatively, if it is to be used for surveying or navigation then the shape must be preserved and the projection has to be conformal.

2. Secondly, a good projection is one that minimises the scale factor over the region; that is, the scale factor must be everywhere as close to unity as possible. This may influence the choice of developable surface. For example, a transverse cylindrical projection is suitable for regions that are longer in their north-south extent than their east-west extent, and conic projections are suitable for mid-latitude regions with a large extent in longitude. In minimising the scale factor over the region, there may be some goal for the maximum allowable scale factor distortion, which may lead to the conclusion that a single projection cannot achieve the desired result and the area must be split up into zones. This adds to the complexity of the situation, and makes it difficult to carry out computations between points in different zones. In such situations, it may be that using a projected coordinate reference system is no longer appropriate.

3. Any additional properties would usually be considered after the preservation of important properties and minimising the range of the scale factor. It may be required, for example, that the appearance of the graticule should be as simple as possible, or that meridians should be parallel to each other. In some circumstances this might be considered more important than scale factor: for example, the Mercator projection is used in navigation because of the parallel meridians, which – due to the conformal nature of the projection – means that loxodromes (lines of constant bearing) plot as straight lines. This is despite the fact that it has a higher scale factor distortion than some other projections methods.

Having selected a suitable projection there is now the need to select a base geographic coordinate reference system to which this will be applied. This dictates the geodetic datum and ellipsoid to be used. The choice will usually be between the system used in the area of interest or WGS 84 as used by the global GPS navigation system.

An example of the thinking involved in the process of selecting a suitable projected coordinate reference system is given in Case Study 6.3.

4

TRANSFORMATIONS

4.1 Introduction

This chapter discusses *transformations* – the change of coordinates between reference systems based on different datums. We begin by looking at some general characteristics of any transformation. We then discuss various transformation methods, characterising them by the type of coordinate reference system upon which they operate. Finally in this chapter we examine issues associated with the derivation and selection of transformations.

Figure 4.1 shows subtypes of coordinate reference systems (CRSs) of especial interest. In this figure, coordinate reference systems are shown in rectangular boxes. Datums are represented through colour in larger rectangles. Two of the datums are geodetic, each of which is associated with four types of coordinate reference system – geocentric, geographic 3D, geographic 2D, and projected. The figure also includes two vertical datums and two engineering datums. The dotted and dashed lines between the coordinate reference systems indicate conversion and transformation methods, with the

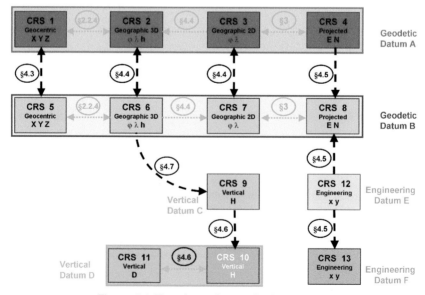

Figure 4.1 Transformation method overview.

number within an ellipse adjacent to the line indicating the section of this book in which the method is described.

4.2 General characteristics of transformations

4.2.1 Transformations and conversions

Mathematical operations for changing coordinates from one coordinate reference system to another may be characterised by whether the operation involves a change of datum. *Conversions* exclude any change of datum, whereas *transformations* include a change of datum.

This distinction between conversion and transformation is useful for emphasising the importance of datums in eliminating ambiguity in coordinates, but unfortunately some mathematical methods such as polynomials can be applied to changing coordinates in different circumstances, some of which include and others exclude a change of datum.

A more useful distinction is whether or not the operation introduces loss of accuracy in coordinates. The method and parameter values of a conversion such as between geocentric Cartesian and geographic coordinates described in section 2.2.4, or the conversion of geographic coordinates to plane Cartesian coordinates through a map projection described in chapter three, are *defined*. As such, they are considered exact. Their application introduces no loss of accuracy in the output coordinates compared to the accuracy of the input coordinates.

In contrast, it will be recalled from Chapter 2 that datums are realised through survey observations and the adjustment of those measurements. Measurements are never perfect and adjustments distribute error throughout the network. A transformation between one datum realisation and another will be affected by the surveying imperfections of both coordinate reference systems.

4.2.2 Transformation multiplicity

The parameter values of a transformation are empirically derived. The coordinates in two reference systems are compared. Because each point will contain slightly different measurement error, if the subset of points used is changed then the derived parameter values will be different. Take for example the heights of coordinates related to different datums given in Table 4.1. Three of the four points have values in both coordinate reference systems and may be used to derive a simple offset transformation.

Table 4.1 Coordinates used to derive a transformation.

Point	Vertical CRS 1 Height (m)	Vertical CRS 2 Height (m)	Difference ΔH (cm)
A	10.20	10.27	+7
B	13.13		
C	13.50	13.55	+5
D	16.30	16.39	+9

Table 4.2 Transformation parameter values for different transformations.

Transformation derived at	Value of height offset (cm)
A only	+7
D only	+9
A and C	+6
A, C and D	+7

Using the average of all three common points A, C and D, the difference in height from CRS 1 to CRS 2 is +7 cm. However, if only points A and C were used to derive the transformation from CRS 1 to CRS 2, the differences would be +6 cm. It can be seen that other choices of points to derive the transformation such as C and D or only D will give yet different values. Table 4.2 shows some of the possible transformation parameter values.

In general, there may be many different transformations between any two coordinate reference systems.

4.2.3 Transformation accuracy

If we accept the transformation derived at points A and C (ΔH = +6 cm) and now apply it to the coordinates of system 1, then the resulting coordinates in system 2 are as given in Table 4.3.

Heights derived through the transformation have an error. If the coordinates in CRS 1 are considered error-free, then applying the transformation will cause the derived CRS 2 coordinates to be less accurate. The average accuracy of this transformation would seem to be about +/– 2 cm. Transformation accuracy will vary with change of location and will be unknown at points such as B, although it can be estimated by using the average at (nearby) known points or by trend fitting.

A word of caution regarding quoted transformation accuracy. The values of transformation parameters are usually derived by least squares techniques. The statistic quoted may be an internal precision rather than an external accuracy. And if the transformation is determined with no redundancy or over a small area in

Table 4.3 Coordinates derived from application of a transformation.

Point	CRS 1 Height (m) (see Table 4.1)	Transformation ΔH applied (m)	CRS 2 Height (m) from transformation	CRS 2 True Height (m) (see Table 4.1)	Error in transformed height (cm)
A	10.20	+0.06	10.26	10.27	−1
B	13.13	+0.06	13.19		
C	13.50	+0.06	13.56	13.55	+1
D	16.30	+0.06	16.36	16.39	−3

which the survey observations are homogonous, a small accuracy value may result. This accuracy estimate may have no validity if the transformation is applied to a wider area. Conversely, as the distortions in the survey networks tend to change slowly, a transformation may be applied over a very local area with little change in relative error.

4.2.4 Transformation reversibility

An important aspect of transformation of coordinates is how the transformation may be reversed and in particular whether parameter values may be used for both forward and reverse calculations. The map projection methods described in chapter three are reversed through the inclusion of different forward and reverse formulae within one conversion method. The projection defining parameters are applicable to both forward and reverse conversions. For the transformation methods described in this chapter, the situation regarding use of the same parameter values in both forward and reverse computations is rather different. Some methods are not reversible: if the reverse computation is required the formulae have to be used with a completely different set of parameter values. Other transformation methods from a theoretical perspective are not exactly reversible, but in the context of their application to geodetic problems may be considered to be reversible for most practical purposes. In yet other methods, the same formulae are applied to forward and reverse calculations and the parameters used for both directions, but some or all of the parameter values have to have their signs changed in the reverse calculation. The importance of this is that it allows a particular transformation – a set of parameter values associated with a method – to be described just once and applied in either direction. But this practicality comes at a cost: care must be taken to identify the direction to which the defining parameter values apply. Throughout this chapter, reversibility is discussed on a case by case basis.

4.3 Transformations between geocentric coordinate reference systems

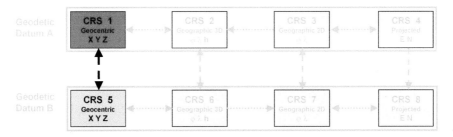

4.3.1 Introduction

Geocentric Cartesian coordinate systems were introduced in section 2.2.3, and it was shown in section 2.3 that the act of introducing a datum makes each one have a different relationship to the Earth. In general, therefore, any one coordinate reference system will require – at least – shifts and rotations in three dimensions, as well as a scale change, in order to transform coordinates referenced to it into any other system.

Transformation methods involving these seven earth-centred parameters are discussed in section 4.3.3. These have some limitations, which can be overcome by using additional parameters, discussed in section 4.3.4. Before that, in section 4.3.2, we examine the simpler three-parameter transformation method. The case study in section 6.1 contrasts the derivation of sets of parameters for these methods.

4.3.2 Three parameter geocentric transformation

Given that each geodetic coordinate reference system, when defined and established, was at least *attempting* to use an international length standard, to make the minor axis of the ellipsoid parallel to the Earth's rotation, and to use an accurate longitude as its origin, it follows that the most significant difference between coordinate reference systems can be accounted for by the different offsets from the Earth's centre of mass. For this reason, a simple three parameter shift – *geocentric translations* – accounts for a large part of the differences in geocentric coordinates, and is often used as a first approximation.

The transformation of coordinates from the source system (X_S, Y_S, Z_S) to the target system (X_T, Y_T, Z_T) is given by:

$$\begin{pmatrix} X \\ Y \\ Z \end{pmatrix}_{TARGET} = \begin{pmatrix} X \\ Y \\ Z \end{pmatrix}_{SOURCE} + \begin{pmatrix} \Delta X \\ \Delta Y \\ \Delta Z \end{pmatrix}_{S\, to\, T} \tag{4.1}$$

where ΔX, ΔY and ΔZ are the geocentric translation parameter values for the translation *from* the source coordinate reference system *to* the target system. These values are the geocentric coordinates of the origin of the source coordinate reference system expressed in the target system. Table 4.4 gives parameter values for Australia between two realisations of the Australian Geodetic Datum (AGD) and the Geodetic Datum of Australia, GDA94 (ICSM 2002), and for Belgium.

Note that the geocentric coordinates X, Y, Z should not be confused with easting, northing and height or with topocentric Cartesian coordinates, which in some coordinate reference systems take the same abbreviations (see section 2.2): the geocentric offsets ΔX, ΔY and ΔZ must be applied in the geocentric domain.

The three-parameter geocentric transformation can be exact at a single point, but in the general case it gets less accurate the larger the area to which a user attempts to apply it. It is therefore necessary to point out that parameter values quoted are averages across the area over which they have been designated – often whole countries – and should be used with caution. For most classically defined and realised

Table 4.4 Geocentric translation parameter values.

Country	Source CRS		Target CRS	Geocentric translation parameter values		
				$\Delta X(m)$	$\Delta Y(m)$	$\Delta Z(m)$
Australia	AGD66	to	GDA94 ($\approx$ WGS 84)	-127.8	-52.3	$+152.9$
Australia	AGD84	to	GDA94 ($\approx$ WGS 84)	-128.5	-53.0	$+153.4$
Belgium	BD72	to	ETRS89 ($\approx$ WGS 84)	-125.8	$+79.9$	-100.5

coordinate reference systems, one would expect errors of perhaps 5 to 10 m or more to result when using such a three-parameter transformation to convert GPS-derived data into a local mapping system. It has the advantage of being a compact transformation, however, and it is how simple hand-held GPS devices store and apply the details of transformations to many national mapping coordinate reference systems.

Reversibility

This transformation method is easily reversible. That is, if parameter values are quoted for the transformation of one system *to* another then equation 4.1 can be simply re-arranged to give the transformation in the opposite sense. But it is more convenient to always use equation 4.1, in which case for the reverse transformation the revised parameter values, such as given in Table 4.5, should be used.

Care is required to ensure that the direction of the transformation is clearly stated and that the parameter values are applied with appropriate sign.

4.3.3 Seven parameter geocentric transformation

The translations required along the three geocentric axes can amount to several hundred metres – perhaps over a kilometre in the most extreme cases. These generally account for the largest changes when transforming from one geodetic coordinate reference system to another. However, the three-translation method does not take into account the misalignment of axes that would result in each system making a different realisation of the direction of the spin axis, nor the effect of the propagation of different length standards. A more accurate transformation can therefore be brought about by taking these phenomena into account and allowing for three-dimensional rotations and scale changes between the coordinate systems. With rotations about each of the three axes, the scale change, and the original three translations, this makes for a seven parameter transformation method.

$$\begin{pmatrix} X \\ Y \\ Z \end{pmatrix}_{TARGET} = \mu \mathbf{R} \begin{pmatrix} X \\ Y \\ Z \end{pmatrix}_{SOURCE} + \begin{pmatrix} \Delta X \\ \Delta Y \\ \Delta Z \end{pmatrix}_{S\,to\,T} \tag{4.2}$$

where

μ is the scale factor between the two systems. For transformations carried out between national coordinate reference systems, this will be a number very close to unity (differing at the level of a few parts per million).

$\mathbf{R}$ is a rotation matrix. The sign convention for the rotations is slightly problematic, in that there are two different ones in general use.

Table 4.5 Transformations from Table 4.4 with direction reversed.

Source CRS		Target CRS	Geocentric translation parameter values		
			ΔX(m)	ΔY(m)	ΔZ(m)
GDA94	to	AGD66	+127.8	+52.3	−152.9
GDA94	to	AGD84	+128.5	+53.0	−153.4
ETRS89	to	BD72	+125.8	−79.9	+100.5

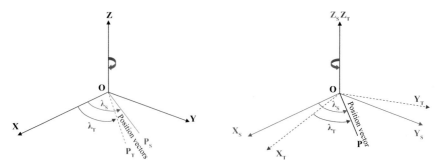

Figure 4.2 **The position vector (left) and coordinate frame (right)**
rotation conventions.

The more common rotation convention, which is now used by the International Association of Geodesy, is termed the *position vector* convention. This defines a positive angle as a clockwise rotation of the position vector about the axis when viewed from the origin of the Cartesian system along the positive direction of the axis. For example, a positive rotation about only the Z-axis from source system to target system α_Z will result in a larger longitude value for the point in the target system.

Using the position vector convention, for small angles α_X, α_Y, α_Z about the X, Y, Z axes, respectively, the rotation matrix can be expressed as:

$$\mathbf{R} = \begin{pmatrix} 1 & -\alpha_Z & \alpha_Y \\ \alpha_Z & 1 & -\alpha_X \\ -\alpha_Y & \alpha_X & 1 \end{pmatrix} \tag{4.3}$$

The alternative convention for the sign of the rotation angles is termed the *coordinate frame rotation* convention. In this a positive rotation is defined as a clockwise rotation of the reference frame (as opposed to the position vector) when viewed from the origin along the positive direction of the relevant axis. A positive rotation about only the Z-axis from source coordinate reference system to target coordinate reference system will result in a smaller longitude value for the point in the target coordinate reference system, as shown in Figure 4.2.

The coordinate frame convention rotation matrix can be expressed as:

$$\mathbf{R} = \begin{pmatrix} 1 & \alpha_Z & -\alpha_Y \\ -\alpha_Z & 1 & \alpha_X \\ \alpha_Y & -\alpha_X & 1 \end{pmatrix} \tag{4.4}$$

A negative rotation in the position vector convention is equivalent to a positive rotation in the coordinate frame convention.

Both position vector and coordinate frame methods are sometimes referred to as Helmert transformations. Both may also be called 'Bursa-Wolf' transformations, but Bursa used only the coordinate frame rotation convention when he demonstrated that, for small angles 'reaching only a few seconds of arc, the three rotation matrices could be combined, an approximation good to better than 1 cm' (Bursa 1966).

The full rotation matrices suitable for large angles take the following form:

$$\mathbf{R} = \mathbf{R}_Z \mathbf{R}_Y \mathbf{R}_X = \begin{pmatrix} \cos\alpha_Z & \sin\alpha_Z & 0 \\ -\sin\alpha_Z & \cos\alpha_Z & 0 \\ 0 & 0 & 1 \end{pmatrix} \begin{pmatrix} \cos\alpha_Y & 0 & -\sin\alpha_Z \\ 0 & 1 & 0 \\ \sin\alpha_Y & 0 & \cos\alpha_Y \end{pmatrix} \begin{pmatrix} 1 & 0 & 0 \\ 0 & \cos\alpha_X & \sin\alpha_X \\ 0 & -\sin\alpha_X & \cos\alpha_X \end{pmatrix} \tag{4.5}$$

Table 4.6 Examples of transformation parameter values using the seven-parameter similarity transformation methods.

	Netherlands	**Belgium**	
From Source CRS	Amersfoort	BD72	
To Target CRS	ETRS89 ($\approx$ WGS 84)	ETRS89 ($\approx$ WGS 84)	
Rotation convention	Coordinate frame	Coordinate frame	Position Vector
Parameter		**Parameter values**	
ΔX	565.2369 m	−106.8686 m	−106.8686 m
ΔY	50.0087 m	52.2978 m	52.2978 m
ΔZ	465.6580 m	−103.7239 m	−103.7239 m
α_X	1.9725 μrad	−0.3366"	0.3366"
α_Y	−1.7004 μrad	0.4570"	−0.4570"
α_Z	9.0677 μrad	−1.8422"	1.8422"
μ	4.0812 ppm	−1.0000012747	−1.0000012747

For the approximation used in the position vector and coordinate frame methods, the angles are assumed to be small and the cosines of the full matrices considered to be unity.

Official transformations such as those shown in Table 4.6 can quote rotations in any units – such as seconds of arc or microradians – but when used in equations 4.3 and 4.4 they must first be converted to radians.

Note that for the Belgium transformations, the values for the three translations differ significantly from those for the 3-parameter method given in Table 4.4. The corollary is that using the three translation subset of a 7-parameter transformation will give significantly erroneous results.

Geocentric 7-parameter transformation software might implement only one of these two rotation conventions. It is important to note that when using quoted parameters they must be used with the model for which they are designed: using parameters with the wrong model will result in a rotation in the opposite sense to that intended. For transformations of geodetic datums, the angles are generally quite small, but when the point of rotation is the centre of mass of the Earth this can represent a substantial error in the resulting coordinates.

Reversibility

These 7-parameter transformation methods use an approximation formula that is valid only when the transformation parameter values are small compared to the magnitude of the geocentric coordinates. Under this condition, for practical purposes the transformation is considered to be reversible. As with the 3-parameter geocentric translation method, the parameter values are applied with their signs reversed for the reverse transformation.

Note that for some extreme cases, such as in Reunion, the rotations are so large that separate non-reversible forward and reverse parameter sets are required. This is usually a strong indication of a poorly conditioned derivation, discussed in the following section.

Applicability

The 7-parameter transformation methods are most commonly encountered when transforming data acquired in a modern system such as GPS to a national coordinate reference system. If the original surveys that established the latter were error free, then the seven-parameter transformation would achieve a good match between the source data and the target coordinate system. In practice, however, there are many more effects besides the overall shifts, scale change, and coordinate misalignments that need to be accounted for. These occur due to gradual accumulations of errors as the original terrestrial surveys were carried out to establish monumented coordinates on the ground, meaning that the target coordinate reference system is distorted and no single similarity transformation will provide a perfect fit. Errors in the order of one, two, or more metres might be encountered, depending on the extent of the area covered and the age of the original survey.

If a better fit is required between newly acquired data and an established coordinate reference system there are essentially two choices. One is to adopt a transformation that introduces a degree of distortion into the geometry of the source data, warping it onto the target system: this type of transformation is covered in section 4.4.4. The other is to split the area up into smaller patches and use different transformation parameters in each one: the smaller the area, the smaller the distortions in the original surveys, and the better the fit that can be obtained. A disadvantage in this latter approach is that discontinuities occur at the boundary between the patches.

Note that these 7-parameter geocentric transformation methods, the 3-parameter geocentric method discussed in the previous section and the 10-parameter geocentric model discussed below are referred to as *similarity transformations*, as they preserve the shape of the original data. Confusingly, the name of this class of mathematical transformation has been adopted in the mapping community as the name for one specific method, discussed in section 4.5.3. In this book, we write the first letter of the name of that particular method in upper case i.e. the Similarity transformation.

4.3.4 Ten parameter geocentric transformation

The 7-parameter geocentric transformation described by equation 4.2 is not the only way of carrying out a three-dimensional transformation, and in many ways it is not the best.

The 7-parameter methods are sensitive to the geometry of the situation. The rotations are about the centre of the ellipsoid. If the area over which points used in the transformation derivation is small, the angle subtended by this area at the rotation point is small and the problem is ill-conditioned. If the problem is ill-conditioned, a rotation around the coordinate origin (at the centre of the ellipsoid, several thousand kilometres away) is very similar to a translation, and a large additional shift may be needed to compensate for this. There is a high degree of correlation between the parameters. As such the solution is not solving for the seven variables.

When derived over small areas, the solution is very sensitive to small errors in the coordinates of the points used in the derivation. A rule of thumb for acceptability is that the subtended angle should be no less than 30 degrees. This is the size of a continent.

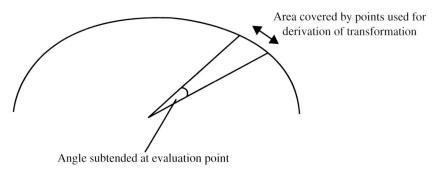

Area covered by points used for derivation of transformation

Angle subtended at evaluation point

Figure 4.3 Derivation of 7-parameter geocentric transformation parameters.

An alternative model is one that makes the rotation about a location at the centre of the points to be transformed, rather than the centre of the geocentric Cartesian coordinate system. The result from applying the transformation will be the same, but the alternative approach has the advantage of making the transformation parameter values so derived easier to interpret. In the case where the rotation is about a more local point, there is very little correlation between the translations and the rotations, and the shifts derived will be much closer to the typical value for the offset between the two coordinate reference systems.

The transformation about a local origin is expressed through the following equation:

$$\begin{pmatrix} X \\ Y \\ Z \end{pmatrix}_{TARGET} = \mu\mathbf{R} \begin{pmatrix} X-X_O \\ Y-Y_O \\ Z-Z_O \end{pmatrix}_{SOURCE} + \begin{pmatrix} \Delta X \\ \Delta Y \\ \Delta Z \end{pmatrix}_{S\,to\,T} + \begin{pmatrix} X_O \\ Y_O \\ Z_O \end{pmatrix} \tag{4.6}$$

where X_O, Y_O and Z_O are the coordinates of a point at the centre of the survey area in the source coordinate reference system, and other terms were defined in section 4.3.2.

When the rotation matrix $\mathbf{R}$ uses the coordinate frame rotation convention of equation 4.4, this method is sometimes known as Molodensky-Badekas.

It needs emphasising that the 10-parameter transformation of equation 4.6 and the 7-parameter transformation of equation 4.2 have exactly the same effect when applied using their correct parameter values. The only difference will be in the appearance of those parameters, as can be seen by contrasting the Dutch transformations in Tables 4.6 and 4.7. The rotations have exactly the same values between the two, but the values of the translations are very different.

Reversibility

The 10-parameter transformation strictly speaking is *not* reversible, i.e. in principle the same parameter values cannot be used to execute the reverse transformation. This is because the evaluation point coordinates X_O, Y_O, Z_O are in the forward direction source coordinate reference system and the rotations have been derived about this point. In principle, they should not be applied about the point having the same coordinate values in the target coordinate reference system, as is required for the reverse transformation.

Table 4.7 Examples of transformation parameter values using the Molodensky-Badekas 10-parameter transformation method.

	Netherlands	**Venezuela**
From Source CRS	Amersfoort	La Canoa (PSAD56)
To Target CRS	ETRS89 (≈ WGS 84)	REGVEN (≈ WGS 84)
Parameter	**Parameter values**	
ΔX	593.0297 m	−270.933 m
ΔY	26.0038 m	115.599 m
ΔZ	478.7534 m	−360.226 m
α_X	1.9725 μrad	−5.266"
α_Y	−1.7004 μrad	−1.238"
α_Z	9.0677 μrad	2.381"
μ	4.0812 ppm	−5.109 ppm
X_O	3903453.148 m	2464351.59 m
Y_O	368135.313 m	−5783466.61 m
Z_O	5012970.305 m	974809.81 m

However, in practical application there are exceptions when applied to the approximation of small differences in the geometry of a set of points in two different coordinate reference systems. The typical vector difference in coordinate values is in the order of $6*10^1$ to $6*10^2$ m, whereas the evaluation point on or near the surface of the earth is $6.3*10^6$ m from the origin of the coordinate systems at the Earth's centre. This difference of four or five orders of magnitude allows the 10-parameter transformation method to be considered reversible for practical application.

Note that in the reverse transformation, only the signs of the translation and rotation parameter values and scale are reversed; the coordinates of the evaluation point remain unchanged.

4.4 Transformations between geographic coordinate reference systems

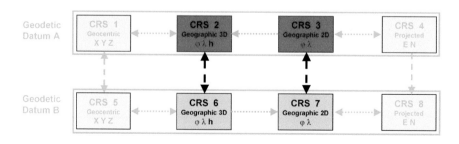

4.4.1 Introduction

We now turn to the transformation of ellipsoidal coordinates referenced to one geodetic datum to values referenced to another datum. Initially we examine formulae for the transformation of 3D ellipsoidal values, before looking at approximate (section 4.4.3) and more exact (section 4.4.4) methods for 2D transformation of latitude and longitude. Finally in section 4.4.5 we introduce the concept of indirect transformation through the concatenation of several steps discussed in earlier parts of the book.

4.4.2 Molodensky and Abridged Molodensky

A set of formulae to transform from one set of ellipsoidal coordinates to another is due to Molodensky (Stansell 1978). Changes in latitude, longitude and ellipsoid height $\Delta\varphi$, $\Delta\lambda$ Δh are to be added to the source coordinates to arrive at the target coordinates:

$$\begin{pmatrix} \varphi \\ \lambda \\ h \end{pmatrix}_{TARGET} = \begin{pmatrix} \varphi \\ \lambda \\ h \end{pmatrix}_{SOURCE} + \begin{pmatrix} \Delta\varphi \\ \Delta\lambda \\ \Delta h \end{pmatrix}_{S\,to\,T} \qquad (4.7)$$

The formulae and additional parameters for computing the changes $\Delta\varphi$, $\Delta\lambda$ Δh are given in Appendix D. The angular units are seconds of arc and the change in height is expressed in metres.

The Molodensky method and transformation parameter values for selected WGS 84 to local transformations are often incorporated into basic hand-held GPS receivers. Some receivers allow the user to enter their own parameter values for Δa, Δf, ΔX, ΔY and ΔZ. Use of this facility is discussed further in sections 4.8, 5.6 and 6.2.

Inspection of the Molodensky equations in Appendix D reveals that the only information required about the nature of the transformation is the change in ellipsoidal parameters and the translations along the three axes. The parameter changes Δa, Δf are the differences that result from subtracting the ellipsoid parameters of the *source* system from those of the *target* system. The terms ΔX, ΔY, ΔZ have the same meaning as when they were introduced in section 4.3 covering geocentric transformations. Therefore these equations are achieving exactly the same result as

- first converting ellipsoidal coordinates to geocentric using equations 2.6 to 2.8 in section 2.2.4;
- then applying the 3-parameter geocentric transformation equation 4.1 of section 4.3.1; and
- finally converting to ellipsoidal coordinates using the parameters of the target ellipsoid through equations 2.9 to 2.11.

This is covered more extensively in section 4.4.5, and it will be seen that this type of transformation offers the increased flexibility of using any of the different types of transformation method at the geocentric to geocentric stage. The only advantage of the Molodensky formulae is in modest computational efficiency, and in the age of the computer it must be questioned how often this is a requirement.

Abridged Molodensky formulae

Should a reader armed with only an abacus require even greater computational efficiency, the Abridged Molodensky formulae given in Appendix D simplify the

Table 4.8 Comparison of full and Abridged Molodensky methods.

	WGS 84	ED50 via full formulae	ED50 via abridged formulae	Difference
Latitude	53°48'33.82"N	53°48'36.565"N	53°48'36.563"N	−0.06 m
Longitude	2°07'46.38"E	2°07'51.477"E	2°07'51.477"E	0.00 m
Ellipsoidal height	73.0 m	28.02 m	28.091 m	0.07 m

formulae to find $\Delta\varphi$, $\Delta\lambda$ and Δh. The difference in results between the full and Abridged formulae for a transformation of a point in the North Sea is shown in Table 4.8.

In the context of a 3-parameter transformation, which has a typical accuracy in the order of 5 to 10 m, these differences are not significant.

Reversibility

Like the 3-parameter geocentric method, both Molodensky and Abridged Molodensky methods are reversible: the same formulae are used but for the reverse case with the signs of the five parameter values reversed.

4.4.3 Geographic offsets

Latitude and longitude values change following the application of a transformation from one coordinate reference system to another, and these changes vary depending on the location of the point being transformed. To what extent, however, might these changes in coordinate be regarded as constant over a fairly limited area? As an example let us consider a test area of approximately 100 × 100 km that is defined by the four points in Table 4.9.

The coordinates in Table 4.9 are defined with reference to a coordinate reference system with the ellipsoid parameters:

$$a = 6378388 \text{ m} \qquad f = 1/297$$

These are converted to a coordinate reference system with the ellipsoid parameters:

$$a = 6378137 \text{ m} \qquad f = 1/298.257$$

Table 4.9 Corner points of a 100 km test square.

Point	Latitude	Longitude
A	50°00'N	00°00'E
B	50°00'N	01°24'E
C	50°54'N	00°00'E
D	50°54'N	01°24'E

Table 4.10 Coordinates in the new coordinate reference system.

Point	Latitude	Longitude
A	49°59'56.29"N	00°00'10.04"E
B	49°59'56.17"N	01°24'09.79"E
C	50°53'56.16"N	00°00'10.23"E
D	50°53'56.04"N	01°24'09.98"E

A 3-parameter geocentric translation transformation is applied, with parameter values of:

$$\Delta X = 200 \text{ m}; \ \Delta Y = 200 \text{ m}; \ \Delta Z = 200 \text{ m}.$$

The resulting coordinates in the new system are shown in Table 4.10. Converting these into *coordinate changes* in metres gives the results shown in Table 4.11.

Although the changes are substantial, it can be seen that the spread is less than 8 m in each direction, and therefore an average 'shift' of the coordinates could be applied with an accuracy of around 4 m. The transformation model would therefore be:

$$\varphi_{TARGET} = \varphi_{SOURCE} + \Delta\varphi$$
$$\lambda_{TARGET} = \lambda_{SOURCE} + \Delta\lambda \qquad (4.8)$$

In the example worked through above, the value of $\Delta\varphi$ would be –3.835", and $\Delta\lambda$ would be 10.01".

This accuracy could, of course, be improved by considering a transformation over a much smaller area, but to approach geodetic or survey accuracy standards special parameters would have to be derived for very small areas indeed.

However, an accuracy of a few metres would be suitable for some leisure (or even commercial) marine users, and therefore this transformation method is one that has been used as an easily published method of converting from satellite navigation systems into the coordinate reference system used on a nautical chart. Nowadays most charts are themselves being transformed onto WGS 84, but on charts of some areas users will see instructions printed in the key that positions derived from satellite

Table 4.11 Coordinate changes caused by transformation.

Point	Latitude	Longitude
A	–115.05 m	200.09 m
B	–118.78 m	195.14 m
C	–118.93 m	200.08 m
D	–122.69 m	195.13 m

navigation equipment should be shifted north/south and east/west by given amounts. This, of course, pre-supposes that the user has not *already* selected the local datum on their navigation system.

It should be emphasised that the parameters published on one chart are only valid at an acceptable accuracy across the area covered by that particular chart, and the same transformation should not be extrapolated to another area.

Reversibility
As with other offset methods, the geographic offset method is easily reversible. As before, the same formulae (equations 4.8) are used but with the signs of the parameters $\Delta\varphi$ and $\Delta\lambda$ reversed.

4.4.4 Grid interpolation – NTv2 and NADCON
A constant theme when dealing with transformations from one geographic coordinate reference system to another is how to handle the problems that arise from distortions caused by the original surveys that were carried out to realise national coordinate reference systems. This is frequently the case when trying to integrate data newly acquired with GPS into national systems. The latter have distortions, but these distortions are part of the coordinate reference system, and to fit in with existing mapping it is necessary for the GPS data to be distorted to fit. One way of dealing with this is to derive transformation parameters for only a small area over which the distortions would not be expected to be too great, and apply a geocentric transformation as detailed in section 4.3. The trouble with this is that everyone ends up using different parameters in overlapping areas, no one is sure where one area changes to another, and so on. One might end up with the national survey organisation, the utilities companies, the local authority and private survey companies all having slightly different transformation parameters in the same area.

One way to overcome this would be to have a universal multi-parameter transformation across the whole of the area covered by the national coordinate system. This implies a considerable effort at the outset, but greatly increased efficiency and enhanced clarity in the future.

The process involves first identifying as many as possible of the monumented control points in the original survey. These are then observed as part of a GPS observation campaign to derive coordinates in WGS 84. This can be done in a hierarchical way: initially the points to be observed can be quite widely spaced out and then a simple similarity transformation derived. Upon examining the vectors of the horizontal residuals of the initial transformation we can identify the spatial correlation: where the vectors change sharply from one control point to another is an indication that the observations need to be densified to detect the underlying pattern of the distortion. This is illustrated in Figure 4.4.

Once a set of common points in the two systems has been acquired at the appropriate density (bearing in mind the accuracy targets), a (usually regular) grid is superimposed on the network. Values for the differences in latitude and longitude are separately interpolated at each of the grid nodes. The techniques for this interpolation vary, but usually involve some kind of surface fitting. It is emphasised that up to this

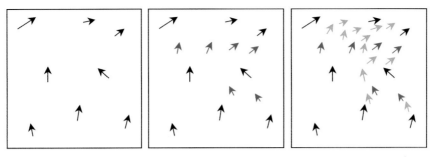

Figure 4.4 **Control points are densified where required until the pattern of residuals is revealed.**

point all the work has been a one-off effort to prepare the transformation, usually by the national mapping authority, and not something that is part of what a user will have to do regularly.

At any point P, the coordinates in the target coordinate reference system are found by applying a correction to its coordinates in the source coordinate reference system

$$\varphi_{TARGET} = \varphi_{SOURCE} + \Delta\varphi$$

$$\lambda_{TARGET} = \lambda_{SOURCE} + \Delta\lambda \qquad (4.9)$$

The corrections $\Delta\varphi_P$ and $\Delta\lambda_P$ are found through bi-linear interpolation (Figure 4.5) of the corrections at the surrounding grid nodes A, B, C and D by:

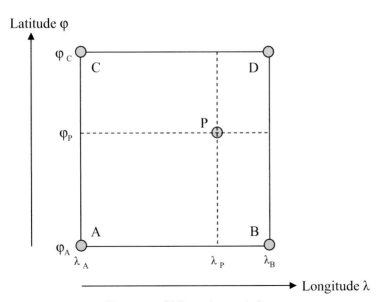

Figure 4.5 **Bi-linear interpolation.**

$$\Delta\varphi_P = \Delta\varphi_A + (\Delta\varphi_B - \Delta\varphi_A)m + (\Delta\varphi_C - \Delta\varphi_A)n + (\Delta\varphi_A - \Delta\varphi_B - \Delta\varphi_C + \Delta\varphi_D)mn \quad (4.10)$$

$$\Delta\lambda_P = \Delta\lambda_A + (\Delta\lambda_B - \Delta\lambda_A)m + (\Delta\lambda_C - \Delta\lambda_A)n + (\Delta\lambda_A - \Delta\lambda_B - \Delta\lambda_C + \Delta\lambda_D)mn \quad (4.11)$$

where

$$m = \frac{\lambda_P - \lambda_A}{\lambda_B - \lambda_A} \quad (4.12)$$

and

$$n = \frac{\varphi_P - \varphi_A}{\varphi_C - \varphi_A} \quad (4.13)$$

The interpolation grids are made available to users by National Mapping Agencies, usually with interpolation software. File formats vary. The NTv2 format, developed in Canada (Junkins and Farley 1995), has also been adopted in Australia, New Zealand and Spain. The NTv2 format includes an estimate of the transformation accuracy at each grid node, and this too can be interpolated at any point using the bi-linear technique. In the US the NADCON format is used for transforming between NAD27 and NAD83; latitude and longitude grids are provided as separate les (Mulcare 2004a). In the UK, the OSTN02 transformation is used for Great Britain (Ordnance Survey 2006).

When transforming between an inhomogenous coordinate reference system determined through classical survey triangulation techniques and a homogenous system that might have been derived through GPS, grid interpolation provides better accuracy than geocentric techniques. Typically, a 3-parameter geocentric transformation might have an accuracy of 5 m, a 7- or 10-parameter transformation an accuracy of 1 m, and bi-linear grid interpolation an accuracy of 0.1 m, depending upon the distortions in the survey networks. However, grid interpolation does not preserve the shape of the original data. Where this is important, a similarity transformation (sections 4.3 and 4.5.3) is more appropriate.

The smoothness of the bi-linear interpolation across grid cell boundaries could be improved through the use of a bi-cubic interpolation using a 4 × 4 array of surrounding points. This additional complexity is rarely necessary for geodetic transformations. The processes of network adjustment and grid derivation tend to produce a smooth surface across which interpolation induces no signicant discontinuity. There would be no increase in accuracy of the transformation.

4.4.5 Indirect transformation between geographic coordinate reference systems

We saw in section 4.4.2 that the Molodensky equations for transforming directly between geographical coordinates are achieving exactly the same result as

- rst converting ellipsoidal coordinates to geocentric Cartesian;
- then applying any one of the 3-parameter geocentric transformation methods of section 4.3.1; and
- nally converting to ellipsoidal coordinates using the parameters of the target ellipsoid.

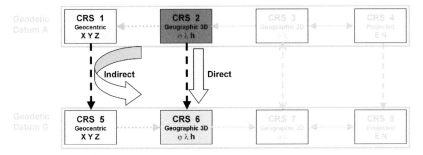

Figure 4.6 Direct and indirect transformation of geographic 3D coordinates.

This procedure can be generalised to use any of the 3-, 7- or 10-parameter geocentric methods for the second step. This transforms from one set of three-dimensional ellipsoidal coordinates (φ, λ, h) to another, as shown in Figure 4.6.

In this process, the geocentric coordinates are always referenced to the Greenwich prime meridian. If the source geographic coordinate reference system is referenced to a different prime meridian, it must first be transformed to the Greenwich equivalent.

$$\lambda_{\text{GREENWICH}} = \lambda_{\text{LOCAL}} + \Delta\lambda \tag{4.14}$$

where $\Delta\lambda$ is the Greenwich longitude of the local prime meridian. If it is the target geographic coordinate reference system that uses a local prime meridian, the transformation is reversed in the now familiar manner of using the same formula (equation 4.14) but changing the sign of the parameter $\Delta\lambda$.

If it is required to transform a set of two-dimensional coordinates (φ, λ) using the same indirect procedure, as shown in Figure 4.7, then it will be seen that on the output side the geographic 2D coordinates are derived from the geographic 3D coordinates by the simple expediency of dropping the ellipsoidal height. But on the input side there is a difficulty in that the geocentric Cartesian coordinates are undefined without knowledge of the ellipsoidal height component.

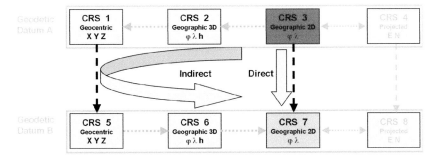

Figure 4.7 Direct and indirect transformation of geographic 2D coordinates.

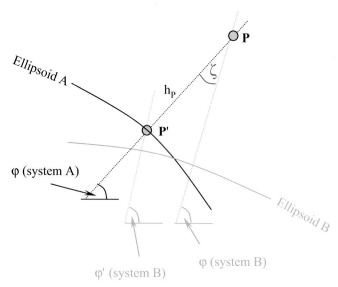

Figure 4.8 **Effect of an error in height.**

However, in practice there is little effect on horizontal position if the ellipsoid height is just assumed to be zero. This point is illustrated in Figure 4.8, in which it is seen that assuming zero height effectively shifts the point P to the point P'. The point P' has the same two-dimensional coordinates in system A as point P, but the two points have different coordinates in system B.

The order of magnitude of this error is given by:

$$\varepsilon = h_p \zeta \qquad (4.15)$$

where

ε is the error in the two-dimensional coordinates (latitude and longitude) in metres

h_p is the ellipsoidal height of point P (in metres)

ζ is the angle between the two normals (in radians).

Since 2000 m is a large value for h_p, and a value for ζ of 20" (arc seconds) is quite large, it is seen that the effect on (φ, λ) of ignoring the height component is of the order of 20 cm. This is significant in some geodetic and surveying applications, but can safely be ignored in the context of transforming mapping data. It would seem safe to assume that two dimensional transformations to a precision of within 0.5 m can always be carried out without a knowledge of height.

If the gravity-related height H_p of the point is known but the correction of this to ellipsoidal height h_p is unknown then, as we have seen in chapter two, the two heights will be no more than 100 m different. Assuming the geoid-ellipsoid separation to be zero and using the gravity-related height as the ellipsoidal height will induce an error in *horizontal* position of no more than 1 cm.

4.5 Transformation of 2D plane coordinates

4.5.1 Introduction

Many users of geospatial data are primarily interested in horizontal location, or require to treat the height component separately. This may be because *only* two-dimensional data has been acquired (perhaps through digitising an historic map or acquiring a single satellite image) or because the height component is less accurate and has separate characteristics. Or it may be because there is insufficient information about the target coordinate reference system for us to be able to adopt more rigorous methods. So although coordinates may be defined formally in relation to a three-dimensional body such as an ellipsoid, we are often presented with two sets of essentially horizontal information and are required to find a simple way of transforming between them.

In the sections that follow, several models for performing such transformations in two dimensions are introduced: they are differentiated by the varying levels of distortion that they impose on the source data to transform it into the target system. The different models are summarised schematically in Figure 4.9.

Typical applications of this type of transformation are:

- A remotely sensed image being warped to fit the coordinates of a set of ground control points.
- GPS data being fitted onto a local site grid with unknown projection or datum parameters. It will not always be apparent to the user that this is what is happening, since the GPS data is 3D and the algorithm will often deal with the height component as well. However, any software package that offers to transform onto a local grid without requiring information on the parameters of its coordinate reference system must be including a 2D transformation of some sort.
- Creating an overlay in a system such as Google Earth™. In some cases this will be quite straightforward; in others users may notice serious discrepancies due to the mis-match between the coordinate system of the map being overlaid and the one implied by Google Earth™.

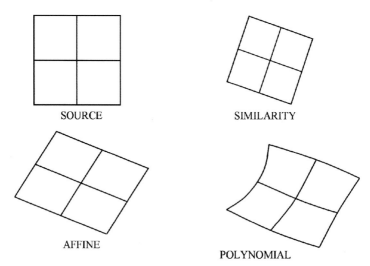

Figure 4.9 Source data set, with the following transformations applied in two dimensions: Similarity (shift, rotation, and scale change); affine (shift, rotation, shear, and different scale changes along each axis); and polynomial (a higher level of warping applied).

- Overlaying a digitised map showing historic land use with a modern data set on a geographic information system. Particular care may have to be taken to account for factors such as paper maps shrinking at different rates in different directions. On the other hand, over-manipulation of the source data may not be desirable since it may mask real changes in the coastline for example.

After the details of the different methods for transformation of 2D plane coordinates have been described in sections 4.5.3 to 4.5.5, two special cases will be considered: creating overlays in systems such as Google Earth™, and transforming GPS data onto a local site grid. These are both important examples of transformation of 2D plane coordinates, but ones in which their use may not immediately be apparent (that is, the user is generally not explicitly offered them as an option). We then complete this section with a look at indirect procedures for changing projected coordinates. However, before then we turn our attention to the general consideration of whether these techniques are *likely* to produce a satisfactory result. That is, the limitations of such an approach will be explored before dealing with the detailed procedures.

4.5.2 Compatibility of coordinate reference systems

It is easy to demonstrate situations where any kind of two-dimensional transformation is not possible by reference to two absurdly different projected coordinate reference systems, both with unknown datum, as in Figure 4.10. No simple linear transformation can change from one system to the other.

We are therefore concerned with a consideration of how close the *shapes* of the two reference systems are. For some situations the answer will be obvious. It might be thought that it will always be possible to transform coordinates from one conformal projected coordinate reference system to another, since by definition the shape is

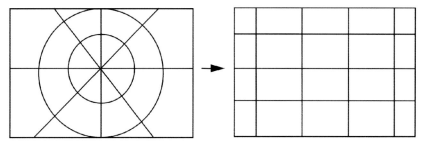

Figure 4.10 Two very different projected coordinate reference systems.

preserved. This condition only holds true for small features, however, since it should be noted that the scale factor varies within the map projection and it is therefore impossible for the shape of a large area to be preserved. An example of this is shown in Figure 4.11, which represents the effect of a variation in scale factor within a map projection on a square of large dimensions.

The point of concern, then, is the variation in the ratio of scale factor between the two map projections across the area concerned. If this figure is largely constant, it indicates that a simple overall scale applied to one image or projection is likely to account for most of the differences between the two.

If, on the other hand, there is a large variation across the area, this gives rise to the question of whether a simple transformation is likely to be adequate.

The key parameter to be considered here is the ratio of the scale factors between the two projected coordinate reference systems, and its variations within the area under consideration. This parameter can be defined as:

$$r = \frac{k_T}{k_S} \tag{4.16}$$

Increasing scale factor

Figure 4.11 Scale factor variations within a map projection.

where k_S is the scale factor in the source system and k_T is the scale factor in the target system, and r therefore represents the scale change in going from one system to the other.

A value of r that was more or less constant across the area concerned would therefore indicate that the two systems have essentially the same geometry, and that a simple two-dimensional similarity transformation is likely to yield satisfactory results.

It is of course necessary to be a little more precise than the statement 'more or less constant'; so consider a situation in which the scale factor ratio across an area ranges from a minimum of r_{min} to a maximum of r_{max}. A similarity transformation applied to the data set as a whole would in effect apply a mean scale factor ratio of r_{mean} (that is, assuming that the control points used to derive the transformation were evenly spread out over the whole area).

The situation might, for example, look like the one depicted in Figure 4.12, in which the scale factor ratio increases from a minimum down the left hand side to a maximum down the right hand side.

If an overall scaling is applied in this example, the right hand side of the area will be scaled along with all other data by the ratio r_{mean}, instead of the correct ratio r_{max}. The mismatch along the right hand side of the image will then be given by:

$$\varepsilon = D\left(\frac{r_{max}}{r_{mean}} - 1\right) \qquad (4.17)$$

where D is in this case the length of the right hand side, or more generally the dimension of the region where the incorrect scale factor is being applied. Equation 4.17 therefore provides a reasonably good rule of thumb for the errors, ε, that are likely to result from the application of the Similarity transformation method (section 4.5.3).

This concept can be extended to a consideration of higher order affine and polynomial transformations, covered in sections 4.5.4 and 4.5.5. These methods can be thought of

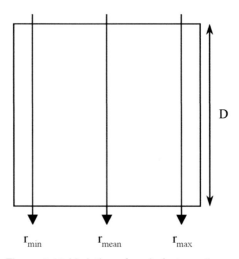

Figure 4.12 Variation of scale factor ratio.

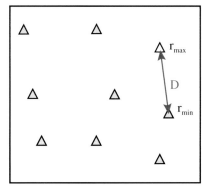

Figure 4.13 **Control for higher order linear transformation.**

in conceptual terms as 'tying' two datasets together at a series of control points that are identifiable in both systems. Exactly the same formula may be applied, but in this situation the extremes r_{min} and r_{max} are to be thought of not across the whole area but between two control points. The dimension D is then the distance between the control points, rather than the dimension of the whole area. This is illustrated in Figure 4.13.

Equation 4.17 will under these circumstances provide a rather conservative estimate of the likely error, as it takes no account of the non-linearity of the transformation. It should be noted, however, that it does assume that a transformation appropriate to the number of control points is being used. That is, there is nothing wrong with using half a dozen control points to derive the parameters of a similarity transformation (in fact, it is laudable), but it should not be assumed that the projection problems have been overcome as a result.

A numerical example of an assessment of the compatibility of two data sets is given in Case Study 6.1.

4.5.3 Similarity transformation method

This 4-parameter transformation method is used to relate a two-dimensional Cartesian coordinate reference system to another two-dimensional rectangular coordinate reference system. It is used when source and target coordinate reference systems

- each have orthogonal axes, and
- each have the same scale along both axes,

for example, between engineering plant grids and projected coordinate reference systems. The Similarity transformation method preserves the internal geometry of the transformed coordinate reference system, so it is ideal for comparing the geometry of any two systems simply by determining the residuals and the root mean square errors after transformation.

By inspection of Figure 4.14, it can be seen that $X_{TP} = X_{TO} + Y_{SP} \sin\alpha + X_{SP} \cos\alpha$ and similarly $Y_{TP} = Y_{TO} + Y_{SP} \cos\alpha - X_{SP} \sin\alpha$.

The figure does not show the relative scaling of the axes of the source and target coordinate reference systems, μ, which also needs to be applied. Dropping the suffix

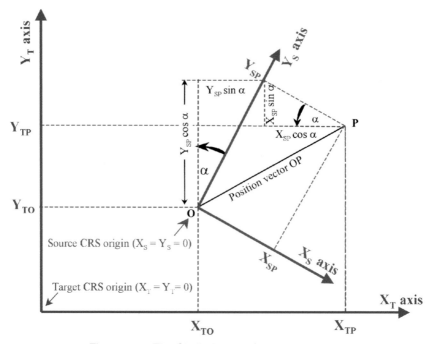

Figure 4.14 The Similarity transformation method.

for point P and rearranging the terms gives the Similarity transformation in geometric form:

$$X_T = X_{TO} + X_S \mu \cos\alpha + Y_S \mu \sin\alpha \qquad (4.18)$$

$$Y_T = Y_{TO} - X_S \mu \sin\alpha + Y_S \mu \cos\alpha \qquad (4.19)$$

where:

$X_{TO}, Y_{TO} =$ the coordinates of the origin point of the source coordinate reference system expressed in the target coordinate reference system;

$\mu =$ the length of one unit in the source coordinate reference system expressed in units of the target coordinate reference system;

$\alpha =$ the angle about which the axes of the source coordinate reference system need to be rotated to coincide with the axes of the target coordinate reference system, counter-clockwise being positive. Alternatively, the bearing of the source coordinate reference system Y_S-axis measured relative to target coordinate reference system north.

Substituting a for $\mu \cos\alpha$, b for $\mu \sin\alpha$, a_0 for X_{TO} and b_0 for Y_{TO}, the Similarity transformation equations 4.18 and 4.19 may be rewritten in parametric form as:

$$X_T = a_0 + aX_S + bY_S \qquad (4.20)$$

$$Y_T = b_0 - bX_S + aY_S \qquad (4.21)$$

Effectively, this transformation is performed by applying

   i) a scale factor, μ, where:

$$\mu = \sqrt{a^2 + b^2} \tag{4.22}$$

   ii) a rotation angle, α, where:

$$\tan\alpha = \frac{b}{a} \tag{4.23}$$

and (iii), two translations (a_0 and b_0).

The operations applied by this transformation (scale, rotation, and translations), are the equivalent of projecting an image by any ordinary photographic enlarger onto a map.

Many mapping software packages will include a facility for deriving the parameters of the transformation if common points can be identified between two data sets. The parameters so derived are then applied to all the other points in the data set to be transformed. With four parameters to be determined, there is a requirement for at least two points of known 2D coordinates. Section 4.9 covers in more detail the requirements for these control or tie points in terms of redundancy, accuracy, and distribution.

Reversibility

The Similarity transformation method is reversible by using alternative transformation parameter values. The reverse parameter values, indicated by a prime ('), can be calculated from those of the forward transformation.

$$X_T = a_0' + a'X_S + b'Y_S \tag{4.24}$$

$$Y_T = b_0' - b'X_S + a'Y_S \tag{4.25}$$

where

$$C = a^2 + b^2 \tag{4.27}$$

$$a' = \frac{a}{C} \tag{4.28}$$

$$b' = -\frac{b}{C} \tag{4.29}$$

$$a_0' = \frac{bb_0 - aa_0}{C} \tag{4.30}$$

and

$$b_0' = -\frac{ba_0 + ab_0}{C} \tag{4.31}$$

4.5.4 Affine transformation

For transformations between two coordinate reference systems where in either system either the scale along the axes differ or the axes are not orthogonal, the affine transformation may be used. The mathematical relationship for an affine transformation in two dimensions may be expressed by the following equations:

$$X_T = a_0 + a_1 X_S + a_2 Y_S \qquad (4.32)$$

$$Y_T = b_0 + b_1 X_S + b_2 Y_S \qquad (4.33)$$

This six-parameter transformation enables scale and rotation adjustments to be applied independently in each direction, and is thus able to correct many effects that have actual physical causes. Thus, for remotely sensed scanner images, it corrects first-order distortions such as affinity due to non-orthogonality and scale difference between scan and along track directions that may be caused by earth rotation and other geometric distortions. For digitised maps, it is able to correct for effects such as a different amount of paper shrinkage in each direction.

At the same time, this kind of transformation may be able to correct for some of the error caused not by physical effects but by differences of datum and map projection between an image and a map, or two different maps. It should be noted, however, that it will not be possible to isolate the magnitudes of physical and non-physical causes.

This transformation applies scale factors in the x direction (or scan direction for a satellite image) of:

$$\mu_x = \sqrt{a_1^2 + b_1^2} \qquad (4.34)$$

and in the y direction (or flight direction for an image) of:

$$\mu_y = \sqrt{a_2^2 + b_2^2} \qquad (4.35)$$

as well as a factor of affinity:

$$F_a = \frac{\mu_x}{\mu_y} \qquad (4.36)$$

As with the Similarity transformation method, the transformation parameters may be derived by identifying points with coordinates that are known in both systems. With six parameters to be determined, a minimum of three 2D tie points must be found. Again, further consideration of the requirements when selecting control or tie points is given in section 4.9.

The Similarity transformation method can be considered as a special case of the affine transformation where parameter $a_1 = b_2 = a$ and $a_2 = -b_1 = b$ (OGP 2007b).

Reversibility
The reverse operation is another affine transformation using the same formulae but with different parameter values. The reverse parameter values, indicated by a prime ('), can be calculated from those of the forward transformation as follows:

$$C = a_1 b_2 - a_2 b_1 \qquad (4.37)$$

$$a_0' = \frac{(a_2 b_0 - b_2 a_0)}{C} \qquad (4.38)$$

$$b_0' = \frac{(b_1 a_0 - a_1 b_0)}{C} \tag{4.39}$$

$$a_1' = + \frac{b_2}{C} \tag{4.40}$$

$$a_2' = - \frac{a_2}{C} \tag{4.41}$$

$$b_1' = - \frac{b_1}{C} \tag{4.42}$$

and

$$b_2' = + \frac{a_1}{C} \tag{4.43}$$

Then equations 4.32 and 4.33 are used with the new parameter values:

$$X_T = a_0' + a_1' X_S + a_2' Y_S \tag{4.44}$$

$$Y_T = b_0' + b_1' X_S + b_2' Y_S \tag{4.45}$$

4.5.5 Polynomials

Second order (12-parameter) polynomials in the form:

$$X_T = a_0 + a_1 X_S + a_2 Y_S + a_3 X_S^2 + a_4 Y_S^2 + a_5 X_S Y_S \tag{4.46}$$

$$Y_T = b_0 + b_1 X_S + b_2 Y_S + b_3 X_S^2 + b_4 Y_S^2 + b_5 X_S Y_S \tag{4.47}$$

are used in remote sensing for the correction of scanner data. If polynomials are used, great care must be taken to ensure that a sufficient number of control points is available and that they are distributed over the whole area to be transformed, as these transformations can behave in an extremely unstable manner.

A minimum of six ground control points are necessarily required to determine the transformation parameters, although it is desirable to have more to build in checks. In addition to first-order distortions, polynomials correct second-order distortions in satellite images caused by pitch and roll, sub satellite track curvature and scan line convergence due to earth rotation and map projection. They may also correct some of the distortions related to the attitude variations along the flight path.

Additional terms may be added to equations 4.46 and 4.47 to correct for higher order distortions; the need for care in the use of control points is greater for higher orders.

This transformation method is also used in geographic information systems to relate data sets that do not match after Similarity or affine transformations have been carried out. The use of the equations is of course invisible to the user; it is simply necessary to identify the minimum number of common points. Again, care must be taken to use points that are well distributed over the area concerned.

Problems can arise if two data sets of very different geometry are being linked in this manner, as the equations can become unstable. In such situations, the user should question whether this type of transformation is really suitable and perhaps at least carry out some initial attempt to re-project one of the data sets into a more compatible coordinate reference system.

4.5.6 Creating overlays in Google Earth™

Google Earth™ is a web based mapping system that displays satellite imagery and map features in an azimuthal perspective projection (section 3.4.7). The *overlays* facility allows users to incorporate additional data layers – for example more detailed maps, historic maps, land use data, and so on. As it is in general designed for non-professional users, a reasonably simple method needs to be provided for registering the new layer into the underlying data structure.

Of course in writing about such a system, we have to bear in mind that web based mapping systems are likely to develop and evolve at a faster rate than amended editions of this book can be published. We are obliged to describe the system as it stands in 2007 and this may not correspond to the situation as a future reader finds it. However, the educated reader should be able to judge any future enhancements to systems such as Google Earth™ using the general principles that have been described in this book.

The system *assumes* that the new layer is in the cylindrical equidistant projection (evenly spaced latitude and longitude intervals). The area selected for insertion of the new data is therefore automatically highlighted as a block between two meridians and two parallels. Since the process is actually viewed as an azimuthal perspective projection, this shows up in the browser not as a rectangle but as a curvilinear area as shown in Figure 4.15. Subsequent to this, the user is permitted to manipulate the overlay with

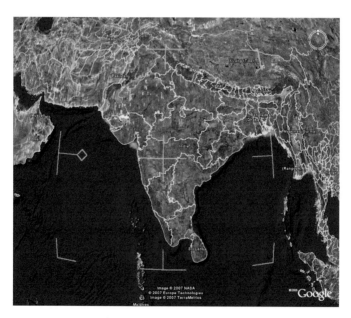

Figure 4.15 Target region for insertion of an overlay in Google Earth™.
(Courtesy of: TerraMetrics, Inc. www.truearth.com
and Europa Technologies Ltd. www.europa-tech.com)

what amounts to a five parameter affine transformation (two translations, rotation, and differential scale along each of the two axes).

What needs to be noted here is that the azimuthal perspective projection is just for viewing, and its geometry has nothing to do with the successful registration of the overlay. What does matter is the relative geometry of the overlay's *actual* projection and the cylindrical equidistant that Google Earth™ expects. At this stage it is convenient to assume (temporarily) that the overlay is in a projected coordinate reference system that has been designed for the area that it depicts, and that its scale factor is everywhere quite close to unity: therefore most of the problems will come from the rapid change in scale factor of the cylindrical equidistant projection.

Reference to section 3.3.1 shows that the cylindrical equidistant projection has a scale factor along the meridians of 1, whilst along the parallels it is equal to $\sec\varphi$. However, because an affine transformation is permitted, the overall differential scale can be taken out by stretching the overlay in the east-west direction. So what remains as a problem is the scale factor ratio on the northern and southern edges between the extremes of scale factor and the average value applied. Let us say that we have an overlay with a range of latitude between φ_{MIN} and φ_{MAX}, and that we apply a re-scaling of $k_{P(MEAN)}$ along the parallels, derived from the average of the extreme values. Then we can derive a figure for the potential mis-match of the registration from:

$$\varepsilon = \frac{D}{2}\left(\frac{2\sec\varphi_{MAX}}{\sec\varphi_{MAX} + \sec\varphi_{MIN}} - 1\right) \qquad (4.48)$$

D is the overall east-west dimension of the overlay (at ground scale) and ε adopts the same units.

If we examine equation 4.48 then what we see is that the error gets worse the larger the overall size of the overlay that we are trying to insert, and the greater the latitude (north or south) of where we are trying to put it. To take a few examples:

- We have a scanned map of New Mexico, covering approximately 31.5°N to 37°N, and representing a width of approximately 550 km. We would expect errors of approximately 10 km at the extremes when we try and overlay it.
- Finland ranges from approximately 60°N to 70°N. It has a width of roughly 400 km, which would imply errors of around 40 km.
- A map of Paris, showing tourist sites within the outer ring road. The latitude range is approximately 48.75°N to 48.95°N, and the dimension is around 25 km. We would expect to see errors of around 25 m.

In the latter example, we see a good instance of the reduction in error when dealing with smaller areas, but it should also be remembered that a map of Paris is actually more likely to have been based on the *French* coordinate system than one specially designed for that city. Therefore, depending on the overlay's location within the national mapping system on which it is based, and the type of projection that has been used, we might get a small additional error on top of those so far estimated (or it might even make the error less). However, it is likely to be smaller by comparison, and so equation 4.48 remains a good rule of thumb.

What can be done if the errors outlined are not acceptable? The overlay could be chopped up into smaller pieces and each one registered separately, but this is likely to look a bit of a mess at the joins. It is not inconceivable that future releases of

Google Earth™ will permit overlays to be entered in projections other than cylindrical equidistant and allow the user to specify from a list what their data format is.

Otherwise, the best way would be to convert the overlay from its projected coordinate reference system into geographic coordinates, and then re-project it into the cylindrical equidistant. A software package that permits the manipulation of raster images would be required, and it would be necessary to enter the parameters of the projected coordinate reference system on which the map is based.

An example of this process is given in Case Study 6.5.

4.5.7 Transformation of GPS data onto a local site grid

Most survey-class GPS processing software will offer a facility for transforming new data onto a local site grid. This is generally because an engineering project may have been set up using a 'flat Earth' assumption and no formal projected coordinate reference system established. An engineering coordinate reference system is established when one point is assigned arbitrary coordinates, and one bearing to another point fixed at an arbitrary value. Provided the extent of the project is reasonably small, it may be that no significant problems arise from the assumption that the Earth is flat.

The problem is then how to integrate data acquired with GPS into this system. To begin with we need to include some points whose coordinates are known in the local system into the GPS observational scheme. But after this, we cannot follow the 'conventional' route of converting both coordinate reference systems to geocentric Cartesian, since the information is not available for the engineering grid on how to do this. Instead, the GPS derived coordinates are split into horizontal and vertical components and these are treated separately: the vertical part is dealt with using the techniques covered in section 4.6.4. 'Horizontal' of course means 'coordinates projected onto the WGS 84 ellipsoid', which will not in general be parallel to the horizontal plane of the local grid system. This has a very minor effect on horizontal position, however.

To be able to carry out the transformation onto the engineering grid, the software first converts the ellipsoidal coordinates into projected ones, so that it is then dealing with two essentially two-dimensional coordinate systems. In other words, the software selects an appropriate projected coordinate reference system. This will be done invisibly to the user; appropriate choices might be a Transverse Mercator with a central meridian through the centre of the survey area and with a central meridian scale factor of unity, or perhaps an azimuthal stereographic with no overall re-scaling. Over small project areas a few kilometres across, it will be very difficult to tell the difference between the two.

Now that the two systems are both in apparently two-dimensional plane coordinate reference systems, the software will derive the parameters of one of the similarity, affine or polynomial transformation methods from a comparison of the common points, and then apply this transformation to all the newly acquired GPS data. Generally the similarity transformation method would be the most appropriate, as this preserves the geometry of the original GPS data. Otherwise, the transformation would be very susceptible to small errors in the original coordinates of the target system.

In principle this procedure works fine for relatively small engineering sites: those of more than a kilometre or two across would in any case probably have had their control established by GPS originally and the question of using this procedure would not

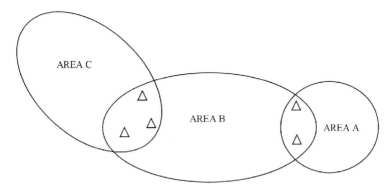

Figure 4.16 Successive development areas of an engineering project grid extended by GPS.

arise. Where it starts to go wrong is when an attempt is made to extrapolate the project grid beyond the area in which it was originally defined. For example, in Figure 4.16 a grid was originally established in Area A, but when it was required to extend the project to Area B this was done by GPS and occupying some local control points. At a later date the project was extended to Area C, again by occupying some local control points.

These extensions have worked quite well in terms of preventing the build up of scale factor as much as it might have done over a large project: this may be a desirable outcome in many engineering schemes. However, it has been done at the expense of any real coherence in the engineering coordinate reference system that has been created. The projection of Area A is undefined, while those for B and C are rotated Transverse Mercator, all with scale factors that are approximately similar in the overlap areas. Where Area B becomes Area C is fuzzily defined, and if the project grew even larger and an Area D was established to the north and including common points with all three of the others, then none of it would make a great deal of sense. Furthermore, the danger with this type of development is that the transformation parameters between WGS 84 and the local site grid (or rather, the different versions of the parameters that must be defined for each area) are difficult to quote to a client in any succinct form, and end up just existing on the laptop of someone, somewhere.

The alternative to this would have been to decide at the point that the project was extended that a local grid was no longer appropriate and design a suitable projected coordinate reference system to cover the whole project, then use one of the two dimensional transformation methods described in sections 4.5.3 to 4.5.5 to go from the old data in Area A into the newly defined system (rather than the other way around).

Another danger with this type of automatic transformation of GPS data is when it is used as a short cut to transform new data onto a national coordinate reference system, rather than a local site grid. The difference here is that the local site grid had very little distortion due to projection effects, whereas the national system is optimised for a much larger area, and therefore the scale factor might be changing much more rapidly. This might have been done because of difficulties in obtaining the parameters of the coordinate reference system being transformed into, or because it seemed much more convenient than researching them. In such situations we can use equation 4.17 to examine the problems that might be caused. As an example, transforming onto local control points spread out over about 5 km at 3° away from the central meridian of a local projected coordinate reference system would lead to mis-matches of around 10 cm. This is likely to be a problem when bearing in mind the potential accuracy of satellite positioning systems such as GPS and therefore should be avoided if at all possible.

That said, some local reference systems have so many distortions due to errors in the old surveys that something like this approach sometimes has to be adopted to counteract them. However, where possible it should only be done over a very small area (less than 1 km across would lead to errors of less than 1 cm due to the projection effects for most practical coordinate reference systems).

4.5.8 Indirect transformations between projected coordinates
We have seen in section 4.4 that a chained indirect transformation through geocentric coordinates may be used as an alternative to a direct transformation between two geographic coordinate reference systems. As an alternative to the transformations of 2D plane coordinates discussed in the sections above, the same chaining principle may be applied to transformations between projected coordinate reference systems, as shown in Figure 4.17. Here there are two cases, with or without a change in geodetic datum.

When a change of datum is involved, such as between CRS 4 and CRS 8, the direct transformation is replaced by three or more steps. The coordinates related to the source projected coordinate reference system may be converted to the base geographic system, a direct transformation such as bi-linear grid interpolation (section 4.4.4) applied to transform the geographic coordinates onto the target datum, and these then converted back into a projected coordinate reference system. The target projected coordinate reference system may use the same projection as the source projected coordinate reference system, but this does not have to be the case. Typical transformations might be:

- from NAD27 / UTM zone 15N to NAD83 / UTM zone 15N
- from NAD27 / California state plane zone VII to NAD83 / California state plane zone V

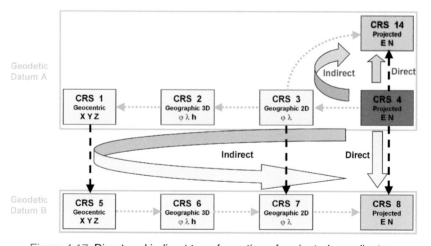

Figure 4.17 Direct and indirect transformation of projected coordinates.

As an alternative to using a transformation between geographic coordinate reference systems, the source map grid coordinates could be transformed through geographic coordinates to geocentric values, and a 3-, 7- or 10-parameter geocentric transformation (section 4.3) applied.

The user typically does not have to be concerned about the multiple steps that a coordinate conversion and transformation application might make. However, he or she should be interested in the transformation being applied as one of the steps. This is discussed further in section 4.8.

When no change of datum is involved, such as between CRS 4 and CRS 14 in Figure 4.17, for example if converting coordinates between two UTM zones such as WGS 84 / UTM zone 31N to WGS 84 / UTM zone 32N, two steps are involved, both conversions. The first converts UTM zone 31 Cartesian coordinates into the WGS 84 geographic coordinate reference system, and the second converts these geographic values into UTM zone 32 grid coordinates. Map projection conversions are discussed in chapter three. With modern computing capabilities this is a straightforward process and much preferred to using a proliferation of direct transformations between projected coordinate reference systems.

There are advantages in using this indirect approach. Where the source and target coordinate reference systems both use conformal projections, conformality is retained. Referring back to section 4.5.2, the application of a higher order linear transformation would not preserve this property. On the other hand, the direct transformation will be significantly more efficient in execution – perhaps two orders of magnitude faster. It is therefore commonly utilised in computer applications. Users should be aware of the effect the approach has on the properties of their dataset, particularly if it is cell-based such as a 3D seismic bin grid or raster image. But the definition of conformality in Chapter 3 must also be considered: shape is preserved *only over small areas*. Over large areas the only way of exactly preserving both grid geometry and spatial location is to re-grid the data in the target coordinate reference system. This is discussed further in the case study in section 6.4.

4.6 Coordinate operations for vertical coordinate reference systems

4.6.1 Introduction

The two categories of height coordinate that were introduced in Chapter 2 were *ellipsoidal heights* (h) and *gravity-related heights* (H). Changing coordinates between these has several possibilities:

- change of one ellipsoidal height to another;
- changing between two gravity-related height systems;
- change between ellipsoidal height and gravity-related height.

These are shown in Figure 4.18.

The first category of height transformation is from one ellipsoidal coordinate system to another. Ellipsoidal height forms one component of a three-dimensional geographic coordinate reference system, and cannot exist on its own. A change of ellipsoidal height is carried out as part of a complete transformation from one geographic

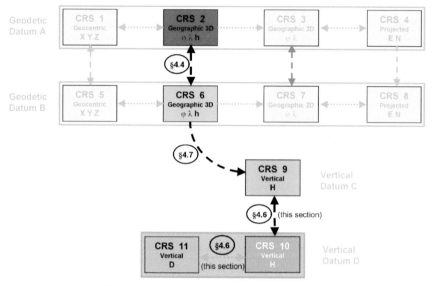

Figure 4.18 Changing vertical coordinates.

three-dimensional coordinate reference system to another and would usually be brought about by one of the methods discussed in section 4.4.

One of the most common types of height transformation – encountered especially when integrating GPS observations with mapping or engineering projects – is between the two different types of height. Because ellipsoidal height is part of a 3D system and gravity-related height is within a 1D vertical system, special considerations apply. These are discussed in section 4.7.

This section concentrates on the change of gravity-related height and depth coordinates between two vertical coordinate reference systems.

4.6.2 Vertical offsets

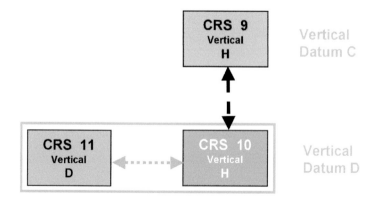

It is possible to convert between two different systems using gravity-related heights or depths. This might be required when there is a change of units, for example from metres to kilometres or to feet. Alternatively, there may be a need to cite depths rather than negative heights. In both cases, there is a change of coordinate system, but no change of vertical datum.

In general it is rare to need to transform between vertical coordinate reference systems based on different vertical datums. Although there are many different height coordinate reference systems in use, they generally exist on different land masses and there are few reasons why they would overlap (traditional spirit levelling could not cross water, and vertical angle observations became too inaccurate over more than a kilometre or two). The exceptions that might be identified are:

- Where a local site datum has been established for an engineering project – perhaps one point was nominated as '100.000 m' pending further information and to allow relative heights to be quoted. At a later stage of the project, it may be required to convert these to a national height coordinate reference system.
- Where there has been an historic change of datum but some values continue to be quoted in the old system.
- Where a single land mass has been politically divided (e.g. Ireland or the continent of Europe) and coordinate reference systems have developed separately, but occasionally projects on the border require transformation between systems.

The first of these is trivial. Once the datum point (or any other in the system) has been connected to the national system an offset will be apparent. Assuming that there are no systematic errors over the local site system, this vertical offset can be applied universally:

$$H_T = H_S + \Delta H \qquad (4.49)$$

The vertical offset transformation method is easily reversible: the same formula (equation 4.49) is used but with the sign of the offset ΔH reversed.

Equation 4.49 is valid only when both source and target systems share the same positive direction of the axis (for heights, upwards but the equation applies equally to two depth systems positive downwards), and when both source and target system share the same units (such as metres). If these conditions are not met then a more general form is given by:

for the forward transformation:

$$X_2 = \{[m * (X_1 * U_1)] + (\Delta A_{1>2} * U_A)\} / U_2 \qquad (4.50a)$$

for the reverse transformation:

$$X_1 = \{m * [(X_2 * U_2) + (-\Delta A_{1>2} * U_A)]\} / U_1 \qquad (4.50b)$$

where

X_1 and X_2 are heights or depths as appropriate and $\Delta A_{1>2}$ is the offset for the forward transformation from source system 1 to target system 2. $\Delta A_{1>2}$ is the value of the origin of the source system 1 in the target system 2

m is a direction multiplier ($m = 1$ if both systems are height or depth; $m = -1$ if one system is height and the other system is depth); and

U_1, U_2, and U_A are unit conversion ratios to metres for systems 1 and 2 and the offset value.

The historic change of datum case is potentially just as straightforward, but the problem is that it is sometimes accompanied by a simultaneous re-levelling campaign. The same vertical offset transformation method may be applied, but now the accuracy of the transformation may become significant. A practical example is in southern Ireland (Eire) where the original levelling was based on a datum point near Dublin called Poolbeg (sometime also referred to as Ordnance Datum Dublin). This was established in the 19th Century and was a measure of the lowest tide over a given period: it became – and remains – the reference to which individual Chart Datums are referenced around the coast by port authorities and the UK Hydrographic Office. Meanwhile, a new tide gauge was established at Malin Head in 1957 and a new determination of mean sea level made, which was propagated through Ireland by a new levelling campaign. This became the vertical datum for gravity-related heights in mapping: it is *approximately* 2.7 m above Poolbeg, but where common points have been identified in both systems the variability has been observed as at least ± 0.10 m (OSI 2000).

Vertical offset by interpolation of gridded data

In some cases the offset ΔH is available through bi-linear interpolation of a gridded data set, for example in the US for converting between NGVD29 and NAVD88 using Vertcon file (Mulcare 2004b). The bi-linear interpolation, which is executed as described in section 4.4.4, clearly needs horizontal position. NGVD29 is associated with NAD27 in a compound coordinate reference system, and NAVD88 with NAD83. Because the difference in NAD27 and NAD83 horizontal coordinate values of a point is insignificant in comparison to the rate of change of height offset, interpolation within the Vertcon gridded data file may be made in either NAD27 or NAD83 horizontal systems.

Vertical offset and slope method

A vertical coordinate reference system is propagated through a *realisation* of the mean sea level surface. Vertical systems with datum origins at significantly different positions will therefore not have a constant offset. In principle the offset is irregular. But in practice the mean sea level surface is relatively smooth over limited distances. It may therefore become practical to model the difference between two vertical coordinate reference systems using an inclined plane.

In Europe, national height systems are related to the pan-European height system EVRF2000 through three transformation parameters and the formula:

$$H_T = H_S + \Delta H + [I_\varphi \rho_0(\varphi - \varphi_0)] + [I_\lambda v_0(\lambda - \lambda_0)\cos\varphi] \tag{4.51}$$

where

I_φ is the value in radians of the slope parameter in the latitude domain, i.e. in the plane of the meridian, derived at an evaluation point with coordinates of φ_0, λ_0.

I_λ is the value in radians of the slope parameter in the longitude domain, i.e. perpendicular to the plane of the meridian.

ρ_0 is the radius of curvature of the meridian at latitude φ_0, and v_0 is the radius of curvature on the prime vertical (i.e. perpendicular to the meridian) at latitude φ_0, described further with formulae in Appendix C.2.

φ, λ are the horizontal coordinates of the point to be transformed, in the ETRS89 coordinate reference system, in radians.

φ_O, λ_O are the coordinates of the evaluation point in the ETRS89 coordinate reference system, in radians, and are parameters of the transformation method.

The horizontal location of the point must always be given in ETRS89 terms. Care is required where compound coordinate reference systems are in use: if the horizontal coordinates of the point are known in the local CRS they must first be transformed to ETRS89 values. These transformations have a typical accuracy of better than 10 cm, with a root mean square value of a few centimetres.

The method is reversible – the signs of the parameter values are to be reversed in the reverse transformation.

4.6.3 The hub concept

The transformation of each national system to the regional system allows national systems to be indirectly related to each other. Coordinates from national system A are transformed into the regional hub system and then into national system B, or vice-versa, as depicted in Figure 4.19. This facilitates international studies, for example through the integration of heights related to the German vertical system DHHN92 with heights from the Dutch system NAP.

This hub concept may also be used for relating horizontal coordinate reference systems, and is a convenient mechanism adopted in many coordinate transformation applications, where the GPS system WGS 84 is adopted as the hub.

To work, it is necessary for there to be a transformation available between the hub system and each of the other systems in the application. This has led to an unfortunate design flaw in some applications, which then require a transformation (to the hub system) as part of a coordinate reference system definition. Coordinate reference systems are self-contained entities, which can exist without knowledge of their relationship to any other coordinate reference system.

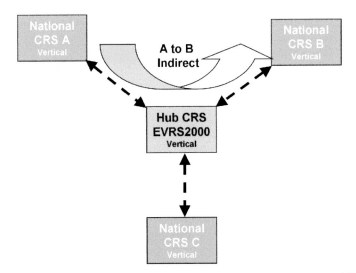

Figure 4.19 The hub concept for relating coordinate reference systems (CRSs) – in this instance vertical systems.

If a direct transformation is available between two coordinate reference systems, then in most circumstances it will be correct to use this rather than go indirectly through a hub system. The selection of transformation is discussed further in section 4.8.

4.7 Transformation between ellipsoidal and gravity-related heights

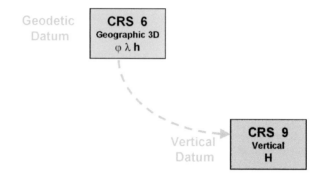

4.7.1 Geoid models

We now turn our attention to the question of how to transform heights in an ellipsoidal system (as would be obtained through GPS observations, for example) to gravity-related heights in a national vertical reference system.

Here we have a difference in cardinality between the coordinate reference systems. Ellipsoidal height is part of a three-dimensional geographic system, whilst gravity-related height forms a one-dimensional vertical coordinate reference system. To retain three dimensions, we need to consider the gravity-related height as part of a compound coordinate reference system (section 2.5). Transformations involving compound coordinate reference systems are discussed further in section 4.7.3. In this section, we concentrate on models that can be used to change ellipsoidal heights to gravity-related heights.

Globally, the geoid may be found from a set of coefficients of a *spherical harmonic expansion*, often referred to as a *global Earth model*. This expresses the geoid over the sphere in terms of a series of functions that are of increasingly small wavelength. The smallest wavelength that the model can express is a function of its highest *degree*: a model that is described as being complete to degree N_{MAX} can model wavelengths down to $180°/N_{MAX}$. A global Earth model used to compute satellite orbits, for example, might be complete to degree 36, which would be sufficient to describe the long wavelength part of the geoid, but not be able to express any effects with a wavelength of less than 5°, or around 500 km.

Until recently, some of the most finely detailed Earth models were produced by Ohio State University, and the models referred to as OSU86, OSU89 and OSU91 (Rapp and Pavlis 1990) were widely circulated and may still be encountered. Being complete to degree 360, these model the form of the geoid down to wavelengths of around 0.5°, or 50 km, and are accurate to about 1 or 2 m. In some parts of the world, the accuracy would be rather worse than this, however, particularly in those areas where a dense network of gravity observations is not available.

OSU91 has now been superseded by a model that has been jointly determined by the US National Geospatial Intelligence Agency (NGA) and the National Aeronautics and Space Administration (NASA), and is referred to as the Earth geopotential model 1996, or EGM96 (Lemoine *et al.* 1998). This too is complete to degree 360, but gains an increased accuracy through the use of additional data, in particular gravity observations that were previously unavailable and new data from satellite missions.

The coefficients of the spherical harmonic expansion of EGM96, together with a computer program for determining the geoid from them, are freely available on the World Wide Web: the location is given in the References under (NGA 2007). Also available is a file of geoid point values determined on a 0.25° grid: a considerable amount of processing is saved if the geoid is interpolated from these values.

Values of the geoid height or geoid-ellipsoid separation N are made available through a *geoid model*. This is a data file from which the value of N can be interpolated for any point. In the OSU and EGM96 models, the horizontal positions of the grid nodes are given in WGS 84 terms. After the separation N has been inter-polated within the model, the gravity-related height is found from a rearrangement of equation 2.15:

$$H = h - N \qquad (4.52)$$

Alternatively, since most precise satellite positioning is done by relative techniques (as described in Chapter 5), equation 4.52 can be re-cast in the form:

$$\Delta H = \Delta h - \Delta N \qquad (4.53)$$

This means that a satellite system such as GPS can be used to find the change in ellipsoidal height (Δh) between, say, a bench mark and a new point. In order to find the gravity-related height of the new point above the bench mark (ΔH), it is necessary to know only the *change* in the geoid separation, ΔN. For low accuracy applications, this can sometimes be assumed to be zero and then $\Delta H \approx \Delta h$. Although in some parts of the world, notably near mountain chains, the geoid-ellipsoid separation may change by 10 m in 50 km, in general ΔN can be considered to be constant over distances of 30–50 km at an accuracy of about 2 m. If a project required centimetre accuracy, it would be unsafe to assume a constant separation over distances of more than 200 m.

Geoid models using additional gravity and height data and therefore of better accuracy locally may be created on a national basis. An example is the AUSGeoid98 model for Australia (Johnston 1998).

4.7.2 Height correction models
It will be recalled from section 2.4.4 that a vertical datum is a *realisation* of the geoid, and therefore not quite synonymous. To change from ellipsoidal height h to gravity-related height H_D based on a specific datum, equation 4.52 may be rewritten as

$$H_D = h - C \qquad (4.54)$$

where C is a value from a height correction surface. The height correction surface differs from the geoid model by small amounts due to the difference between the geoid and the actual vertical datum surface, as well as some other assumptions regarding the gravity field. An example of a height correction surface is the OSGM02 model for relating ETRS89 ellipsoidal heights to Ordnance Datum Newlyn elevations in Great Britain (Iliffe *et al.* 2003). These models have a weak dependence upon horizontal position: the rate of change of height difference over distance is slow. As a consequence, the local horizontal coordinate reference system may be used to interpolate within the model with no significant error in resulting height correction C.

Where no height correction model exists, a global geoid model can be taken as an approximation at the several decimetre level for the transformation of ellipsoid height to gravity-related height.

A recent development is the use of height correction surfaces in hydrographic applications. These facilitate the correction of GPS-derived ellipsoidal heights to Chart Datum. Their advantage is that they eliminate the need to apply tide corrections to observed water depth data in hydrographic surveying operations, the observed depth being applied to the ellipsoidal height (with correction for offset of echo-sounder transducer from GPS antenna), and also allow mariners equipped with a Differential GPS receiver to assess under-keel clearance without knowledge of the state of the tide, thereby eliminating any uncertainty associated with using predicted water levels (FIG 2006). GPS is outlined in Chapter 5.

Height correction models can also be used to relate two gravity-related height systems. An example is that used in France to transform Lallemand height to IGN69 height.

To reverse the application of a geoid or other height correction model, that is to determine an ellipsoidal height from a gravity-related height, requires knowledge of horizontal position for interpolation within the model. This can only be accomplished if the gravity-related height is part of a compound coordinate reference system, discussed in the next section.

4.7.3 Transformations involving compound coordinate reference systems (CRSs)

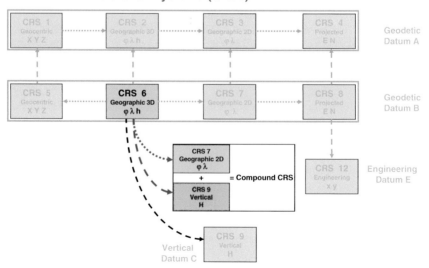

We have seen in section 2.4 that there is a frequent need to describe position in three dimensions where the vertical is a gravity-related height. This one-dimensional vertical coordinate reference system is complemented with a separate, independent, horizontal system. The horizontal system may be truly two-dimensional or it may be the horizontal component of a three-dimensional geographic CRS including ellipsoidal height, as would be obtained through GPS observations, for example. The horizontal 2D system and the 1D gravity-related height system when combined are known as a compound coordinate reference system, as shown diagrammatically in the figure above.

Transforming from the 3D geographic coordinate reference system to a compound system is a straightforward task. It requires splitting the 3D system into horizontal and vertical components and treating them separately. The latitude and longitude may, if required, be transformed to another geodetic datum and/or converted to projected coordinates or an engineering grid through any of the direct or indirect methods described in chapter three and sections 4.3 to 4.5 of this chapter. The vertical component is treated as described in the previous section, 4.7.2.

Reversibility

The reverse transformation from the horizontal component of the compound system to the latitude and longitude component of the geographic 3D system may involve multiple steps but is otherwise straightforward, using any of the direct or indirect methods previously described.

The reverse transformation from the vertical component of the compound system to the ellipsoidal height component of the geographic 3D system requires interpolation within the geoid or height correction model. However, the latitude and longitude arguments for this interpolation must be in the geographic 3D coordinate reference system, as the nodes for the gridded data will be in this system. Therefore the reverse operation on the horizontal component of the compound system must be executed before the reverse vertical transformation can be made.

4.8 Selecting a transformation

4.8.1 Introduction

It was noted in section 4.2 that there may be many different transformations between any two coordinate reference systems. Unlike map projection parameters, which are defined, transformations are derived empirically to best model the distortions inherent in survey networks. They are therefore not exact. Their application introduces error into positions that did not exist before the transformation was applied. Transformations may have been derived for specific purposes or be optimised over limited areas. Transformation methods differ in their ability to model the distortions in the survey data. Choosing one transformation from a multiplicity involves the assessment and balancing of several criteria. There may not be a 'one solution fits all' answer to the problem. This section looks at the issues to be considered. It is assumed that the reader knows the source coordinate reference system to which the data is currently related and the target to which the data is to be transformed.

If a set of transformation parameter values is published for a particular transformation method, then it is up to the user to apply these in the correct way, using the models that are described in the preceding sections. Particular care is required with the sign of the parameter values: if these are incorrect then the effect is to apply the transformation in the wrong direction.

4.8.2 Officially sanctioned transformations

In many countries the national mapping agency will promulgate a national transformation between its historic geodetic coordinate reference system and a realisation of the ITRF (section 2.3.2). This will have been the result of an exhaustive study of the relationship between the two systems over the full area of responsibility of the authority. It represents the best transformation for the country as a whole. Examples include the OSTN02 transformation for Great Britain, the NTv2 method and data files for Canada, Australia, New Zealand and Spain, the NADCON method and data files for the US, and parameter sets for readily available transformation methods such as the geocentric transformations for Belgium and The Netherlands given as examples in section 4.3. A transformation service or downloadable product may be available on-line on the national mapping agency web site.

Adoption of the national transformation provides a consistent standard for mapping and other location-based activities. Users should have a good reason to use an alternative to the officially sanctioned transformation. Reasons might be:

- An alternative, often less accurate, transformation was in use before the publication of the officially-sanctioned transformation and retaining consistency with earlier work is more important than both the improved accuracy from this newer national transformation and consistency with third parties who are using the national transformation. An example might be at the site of an offshore pipeline landfall at which the pipeline operator has determined his own transformation. To allow comparison of annual inspection surveys, the operator prefers to consistently use his own transformation.
- For the purpose required, an alternative transformation is mandated. This may be the case with offshore work where the minerals regulatory authority has decreed a particular transformation be utilised.
- The user's application cannot cope with the transformation method. In the early days of computing, using a simple transformation method such as the 3-parameter geocentric method rather than a resource-intensive bi-linear grid interpolation may have been justified. But with modern computing power this is a rather poor reason. Users should endeavour to use a more suitable application, or have their application vendor add the national transformation to the application's functionality. However, should this not be possible, Case Study 6.2 provides a work-around by deriving parameter values for a simple method from a more complex method.
- A simpler method gives adequate accuracy. This may have been a valid reason when computing power was limited, but is becoming increasingly irrelevant. One area where it retains its validity is in the use of simple hand-held GPS data collection. The accuracy of the position fixing may be 10–15 m or worse, and a 5-m accuracy 3-parameter geocentric transformation will not significantly degrade the accuracy of the data. This is discussed further in Chapter 5.
- The user is working outside the area of applicability of the national transformation. This may be the case when working offshore.

4.8.3 Selecting from a transformation repository

Several resources that provide transformation parameters are available. In some cases the sources are repositories of coordinate reference system definitions, which also hold transformation data; use of these is outlined in section 2.5.2. Many national mapping agencies have their own web sites. In Europe, some of this information is consolidated in the EuroGeographics site maintained on their behalf by BKG (BKG 2007), although this site may not have the latest information available from a national source. The US National Geospatial-Intelligence Agency (NGA) through its unclassified technical report TR8350.2 provides transformation parameter values from numerous local systems to WGS 84 (NGA 2001), and a slightly different set (many of which are built into GPS receiver software) through its technical report TR8358.1 (NGA 1990). These and other transformations have been consolidated in the EPSG Geodetic Parameter Dataset (OGP 2007a). How should one select the most appropriate transformation from several variants?

The first thing to note is that some transformation methods, including those most frequently encountered, are reversible; that is the same formula can be used for converting from system A to system B and from B to A, but with some or all of the transformation parameters applied with the opposite sign. In these cases a transformation repository will normally store the transformation only once and leave it to the user or the user's application to reverse the signs. The consequence of this is that if seeking a transformation from, for example, WGS 84 to ED50, it is necessary to search for transformations both from WGS 84 to ED50 and from ED50 to WGS 84.

Next, many repositories do not delete superseded data. This is to allow users to refer to the information for historic purposes. In some cases the repository will retain erroneous records, but mark them as being deprecated. Unless there is a need to rework coordinates that have already been subjected to a deprecated or otherwise superseded transformation, all transformations so marked should be ignored.

Having eliminated invalid information from the search, the criteria that now should be investigated are:

- The source of the transformation. As discussed in section 4.8.2, there should be a preference in favour of officially sanctioned national transformations.
- The area of applicability for the transformation. Any transformations whose area of applicability does not overlap with your area of interest should be ignored. This should reduce the number of candidates significantly.
- The scope or usage for the transformation. Some transformations have been created for a specific purpose such as regional environmental data management. They may not be appropriate for other purposes such as precision engineering. In some instances there may be a requirement, sometimes legal, to use a particular transformation.
- The temporal validity of the transformation. Has it been replaced by a better determination?
- The accuracy of the transformation. Transformations involving grid interpolation methods generally give better accuracy than geocentric methods; 7- or 10-parameter geocentric transformations will generally give better accuracy than 3-parameter transformations. Can you tolerate the additional inaccuracy this transformation will introduce?
- The transformation method. Does your application have the functionality to use the transformation? Does one method offer significant speed in execution over another? Grid interpolation methods generally are slower than geocentric transformation methods.

If after eliminating unsuitable transformations you are still left with more than one, using all of the information available a choice in favour of the most appropriate will have to be made.

A rather more difficult problem to address is what to do if you are left with no suitable transformation. One solution is to derive your own transformation, following the hints given in section 4.9.

The principles used when selecting from a repository also apply when selecting from a list within an application. But now the selection may be compromised by a lack of ancillary information describing the applicability and accuracy of listed transformations. Remember that many lists hold transformations designed for medium scale mapping with accuracies in the order of 10 m. It may be that none of the transformations listed is appropriate for your work.

In some circumstances you may have no alternative but to use an unsatisfactory transformation in the knowledge that there are significant limitations in its use. In such circumstances it is essential to document what has been done and to retain an audit trail so that the results can be reworked when more suitable data becomes available.

4.9 Deriving your own transformation

4.9.1 Introduction

Deriving a transformation requires resources. It involves acquiring coordinates of control points in both the coordinate reference system in which data is (to be) held, and the system in which the data is required. Before deriving a transformation, it is worth evaluating whether anything existing will suffice. There may then be advantage in consistency with earlier work in using the older transformation; a proliferation of similar transformations simply causes confusion. Testing an existing transformation for applicability in the area of interest will involve obtaining control point coordinates, but does not require the derivation of a new transformation.

The issues examined in this section are firstly the choice of transformation method to be adopted for a new transformation, secondly the availability and sources of control points, thirdly the geometry of those control points, and lastly the effect of ignoring geoid-ellipsoid separation in the derivation of a geocentric transformation. Case Study 6.1 amplifies these discussions with examples of the issues discussed.

4.9.2 Choice of transformation method

The earlier sections in this chapter described a number of transformation methods. Which of these should be adopted?

The first issue to consider is which transformation method best models the distortion characteristics of the two coordinate reference systems. The problem is essentially one of how closely these distortions should be followed when transforming. To begin with, it is helpful to consider what the final aim is; that is, the use that will be made of the coordinate information that is derived. This has a particular bearing in situations where GPS data is being transformed into a coordinate reference system that contains the type of distortions discussed in section 4.5.

To take an example, it is usually the case that an engineering project (such as the construction of a road or railway) should use coordinates that are in sympathy with a local coordinate reference system. But it is quite important that the internal geometry of the project should be preserved: it is not acceptable to have kinks in the railway where the local coordinate reference system is distorted, and it is usually helpful if measurements carried out with ground survey equipment can build onto the GPS control without a continuous need for distortions to be introduced. Here we require a transformation method that will retain the geometry of the GPS data; a similarity transformation – either the Similarity transformation method discussed in section 4.5.3 or one of the geocentric methods discussed in section 4.3 – is likely to be most suitable. The note of caution given in section 4.3.3 regarding the derivation of 7-parameter transformations over small areas must be emphasised here: it is very susceptible to small errors in the coordinates of the control point coordinates in one or both reference systems.

To take a counter example, if GPS is being used to locate features whose coordinates are given in the local system (perhaps underground utilities, for example) then transforming precisely into the local system – with all its distortions – is of paramount importance. Now a transformation method capable of modelling the distortion – grid interpolation or a polynomial – would be appropriate.

A second issue is the software that is available, both for the derivation of the transformation and its later application. These may determine the selection of a particular method. Where the relationship between the source and target systems can be modelled by known geometric characteristics, for example by three offsets and a scale change, the four parameters may be determined as a 7-parameter geocentric transformation with the three rotations constrained to be zero. The Similarity transformation method is a special case of the affine transformation in which the axes of both of the source and target coordinate reference system are orthogonal (that is, Cartesian) and of the same scale, and could therefore be derived using software designed to determine an affine transformation.

4.9.3 Availability of control points

Parameter values for all the transformation methods described in this chapter can be derived provided that a sufficient number of points is available with coordinates known in both the source and the target systems.

What constitutes a control point? This is driven by accuracy requirements. The transformation cannot be more accurate than the control used in its derivation. For engineering projects, monuments surveyed accurately in both coordinate reference systems should be used. For lower accuracy applications, scaling the coordinates of identifiable features from maps or images may suffice. Or in deriving the transformation to fit one map onto another they might simply be points that can be clearly identified in both maps. When geo-referencing a satellite image, they might be specially commissioned surveys with GPS receivers to determine the coordinates of points such as road intersections or building corners that can be clearly seen on the satellite image, or they could be similar points read off a map that is in the required coordinate system. Maps should be used with caution for determining ground control points. Map data at scales of 1:25 000 and smaller is notoriously unreliable because of the many errors that may have accumulated in the map production and map digitising processes. These include: survey errors (in some parts of the world published maps may be based on topographical sketches); drafting errors; generalisation; paper distortion; and errors in digitising a paper document for use in the validation process. It is always necessary to take the accuracy of the data into account when using ground control; it is particularly important when using map data.

The number of points that is *sufficient* will depend upon the number of parameters that it is required to find, and thus on the nature of the transformation: one dimension of a control point is required for each parameter. So, for example, to derive the parameters of a three parameter geocentric transformation, one point known in three dimensions is required. To derive the values of the seven parameters of a similarity transformation, seven pieces of information are needed. This could in theory be provided by two control points that are known in three dimensions and one bench mark (known in height only). The two-dimensional Similarity transformation method needs at least two points to determine its four parameters, and so on.

In fact, the more control points that are available the better. Once the parameter values have been derived, it is then possible to apply these to the known points and note the agreement between the original coordinates and the transformed values. If there is no redundancy then the fit will be perfect, even if there was an error in the coordinates given for the known points. Such an error could occur through incorrect keying-in of the data; alternative sources, for example when fitting GPS data to a local coordinate reference system, are that the control point could have moved (for example through local subsidence) since the last publication of its coordinates, or even that the monument has been demolished and re-built in a slightly different location. In effect, this would mean that in an effort to fit the GPS data to the distortions of the local system, *additional* distortions have been introduced that are not justified.

So it is generally desirable to have more than the minimum number of control points in order to provide a check for errors. With a redundancy of control information (more than the minimum number necessary), it is then possible to derive the optimum transformation parameters by least squares. The mathematical treatment of this subject is dealt with in Appendix E. Once the parameters have been determined, the transformation can be applied to the source coordinates of the control points to see how well they match up with the target coordinates: the differences are termed the *residuals*, and these can be examined to check for any gross errors or unusual patterns. Even then, care must be taken in interpreting the results: with only a slight redundancy (for example using three points in 3D to determine seven parameters) there would be a tendency for the residuals to be much smaller than the size of any errors present. The transformation would stretch things a bit, rotate a bit more, and generally do all it could to fit to the given values: an example of this happening is given in Case Study 6.1.

4.9.4 Geometric issues

As well as having a redundancy of control points, care must be taken with their placement across the area of interest. There are some obvious failure cases: for example, three known points in 3D are necessary for a 7-parameter transformation, but they will not suffice if they all lie in a straight line – the rotation will be indeterminate about the axis that they define. In fact, this example is just a special case of a more general point, that the control points must be well spread around the area in which the transformation is to be applied. Examples of good and bad geometry of control points are given in Figure 4.20.

Circles are data points in the source system that are to be transformed; triangles are control points with coordinates known in source and target systems. The point about the arrangements shown is that slight errors in the coordinates of the control points are 'controllable' when in between the control points, but their effect on transformed data will be greatly magnified the further away one tries to extrapolate the use of the parameters.

For situations where the internal geometry of the project is of more importance than sympathy with existing mapping, relatively few control points are needed. For

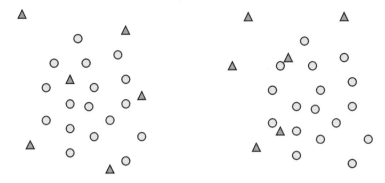

Figure 4.20 **Examples of good (left) and bad (right) geometry of control points. Control points are shown by triangles.**

the opposite case, a large number of control points, well distributed across the survey area, will be necessary.

4.9.5 Effect of ignoring geoid-ellipsoid separation

Transformation methods involving geocentric or geographic coordinates require that the heights used be ellipsoidal. This may require reducing gravity-related heights to the ellipsoid using the models described in section 4.6.4 and the procedures given in section 4.7.

This section considers the alternative of when no information is available on the geoid separation, and yet it is necessary to derive a transformation that will transform GPS WGS 84 data to the local system. Functionally this will proceed in exactly the same way as transforming from one three dimensional geographic system to another, but the gravity-related heights will be 'assumed' to represent heights above the ellipsoid. What effect will this assumption have upon the outcome?

By way of illustration, a hypothetical data set will be used with different geoid characteristics. The points used are shown in Figure 4.21.

In the figure, points A, B, C, and D are known. The coordinates of P are also known, but will be treated as unknown in order to test the quality of the transformation.

The four known points cover an area of 25 km². It is assumed that the local coordinate reference system is parallel to WGS 84, but translated by 100 m in each dimension, and that there are no distortions present in the local system. Such a paragon

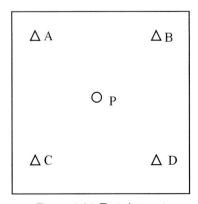

Figure 4.21 **Test data set.**

Table 4.12 Coordinates of test points.

Point	E	N	h	H for case 1
A	400 000.00	125 000.00	50.000	40.000
B	425 000.00	125 000.00	50.000	40.000
C	400 000.00	100 000.00	50.000	40.000
D	425 000.00	100 000.00	50.000	40.000
P	412 500.00	112 500.00	50.000	40.000

of perfection is unlikely, and is only used here to isolate the effect of the geoid. For a more realistic example, in which several sources of error are present simultaneously, see Case Study 6.1.

If the geoid-ellipsoid separation is unknown, then the only option is to take it as zero, and assume that the ellipsoidal heights are the same as the gravity-related heights.

The first case to consider is the one where there is a uniform separation of the geoid across the whole area. In this situation, the coordinates of the control points will all be shifted. For a sufficiently small area (with a uniform value of the geoid-ellipsoid separation), they will all be shifted by the same amount and in the same direction.

A 7- or 10-parameter geocentric transformation can cope with a uniform shift as this will be included in the translations that are derived. (As discussed in section 4.3, over small areas the 10-parameter method is preferred, but to illustrate the issues the 7-parameter coordinate frame rotation method is used here.) What does cause a problem is a change of shape of the distribution of control points. The extent to which this happens is a function of the maximum angle between the shifts caused at the different control points, which is the same as the angle between the ellipsoidal normals. A general rule of thumb for the relative coordinate shift caused by this effect is:

$$\varepsilon = \frac{D}{6400} N \qquad (4.55)$$

where ε is the resulting error, D is the extent of the survey in kilometres, and N is the approximate size of the separation that has been ignored (and has the same units as ε). For a survey of 25 km extent, and a separation of 10 m, this amounts to around 0.04 m. This is significant at the required accuracy level, although a certain amount of this effect is absorbed by the scale factor of the transformation.

To illustrate this, let us assume that the geoid-ellipsoid separation is a uniform 10 m in the test area. Thus, although the heights are all 50 m above the ellipsoid, they will actually be 40 m above the geoid. If the transformation algorithm then assumes (incorrectly) that these heights of 40 m are in fact ellipsoidal heights, the following transformation parameter values are found:

Table 4.13 Transformation parameter values for uniform geoid shift.

ΔX	− 106.302 m
ΔY	− 99.799 m
ΔZ	−107.762 m
μ	− 1.57 ppm

The values of the rotations have not been quoted in Table 4.13, as they are all nearly zero. Clearly these parameter values are 'wrong' in the sense that the translation is known to be 100 m in each dimension, and the scale should be true. What has happened, however, is that the extra 10 m shift caused by the geoid has been absorbed into the transformation. Applying these parameters to the whole data set results in coordinates for P of:

Table 4.14 Coordinates of point P after transformation.

Point	E	N	h
P	412 500.000	112 500.000	40.000

In other words, comparing these coordinates with those in Table 4.12, the plan coordinates are correct to a millimetre and the ellipsoidal height that the program thinks it has found for P is the correct gravity-related height.

As well as being able to cope with a uniform shift of the geoid, a similarity transformation is also able to deal with a uniform tilt, by adjusting the rotations determined during the transformation. This can be illustrated in the test data set by assuming that the geoid slopes as a plane from a height of 10 m on the western side (in line with A and C) to 11 m on the eastern side (in line with B and D). Therefore the levelled heights of the four control points will now be:

Table 4.15 Gravity-related heights for uniform slope of the geoid.

Point	H for case 2
A	40.000
B	39.000
C	40.000
D	39.000

Applying the same procedure as before results in transformation parameter values of:

Table 4.16 Transformation parameter values for uniform geoid tilt.

ΔX	-106.617 m
ΔY	-99.790 m
ΔZ	-108.150 m
α_X	6.40"
α_Y	-0.22"
α_Z	-5.20"
μ	-1.65 ppm

Although the other parameter values are much the same as before (Table 4.13), the rotations are now significantly different, as the coordinates rotate to adapt to the sloping geoid. Applying these parameter values to the data set once again results in:

Table 4.17 Coordinates of point P after transformation.

Point	E	N	h
P	412 500.000	112 500.000	39.500

The plan coordinates are again correct within 1 mm, and the height is also the expected value, as a uniform slope of the geoid would result in a separation of 10.5 m at P.

The 7-parameter geocentric transformation is thus very adept at dealing with uniform slopes and shifts of the geoid. This will quite often suffice for surveys that cover just a few kilometres and are not seeking the highest accuracy.

What a similarity transformation *cannot* cope with is non-uniform changes in the geoid: perhaps a bulge or smaller undulations. In the previous example, as the point P was not included in the derivation of the transformation parameter values, any value of the separation other than that implied by a uniform tilt and shift of the geoid would not have been corrected by this procedure. But the warning in section 4.3.4 should be heeded.

If the undulations of the geoid are sufficiently large to cause a problem at the accuracy required, and no model of the geoid is available, then the only solution to this problem is to incorporate more information into the transformation and to alter the model used. It is possible, for example, that while no extra triangulation points are available in the area surveyed, there may be several bench marks. If these points are included in the GPS survey then a comparison may be made of the heights derived by a similarity transformation with the original bench mark heights. The differences between these two figures will not represent the geoid-ellipsoid separation itself, but in the absence of observational errors will represent a residual value of the separation over and above the overall shift and tilt.

If the residuals so found are indeed due to finer undulations of the geoid, then the expectation would be that they are highly correlated, certainly over short distances. A random spread of residuals with no apparent pattern is more an indication of poor quality GPS data or bench mark heights than short wavelength undulations of the geoid (this is particularly the case in low-lying terrain).

In the situation where the residuals do display a pattern, it is possible to interpolate the geoid values between bench marks to obtain gravity-related heights at the points newly surveyed by GPS.

4.9.6 Evaluating results of the transformation

Once a transformation – the method and its parameter values – has been derived, the statistics available should be examined. These may indicate a problem with some of the coordinates used in the derivation. Where this problem is encountered, an alternative transformation should be derived, with the problem coordinates omitted. Case Study 6.1 gives some examples.

5

GLOBAL NAVIGATION SATELLITE SYSTEMS

5.1 Introduction

For many users, the idea that spatial data can be on different coordinate reference systems, and indeed on different ones to the method of data acquisition, will first have been encountered when using a global navigation satellite system (GNSS) such as GPS. For this reason, although it is possible (and quite reasonable) to write an entire book devoted to these systems, it is appropriate to summarise their most important features here, and to refer the reader to the sections of this book that cover appropriate coordinate reference systems and transformations.

A more extensive treatment can be found in standard texts such as Hofmann-Wellenhof *et al.* (2001) or Leick (2004).

5.2 The systems

Since it was first established as a useable system in the late 1980s, the Global Positioning System (GPS) has for many users been *the* satellite navigation system. Now, however, we are on the threshold of an era in which GPS is to become just one of two, three, or even more different systems.

The introduction of the European Galileo system, the Russian Glonass, and possibly others, together with associated augmentation systems and improvements to GPS itself, means that there is potentially a great deal of information to be absorbed concerning the nature of all the possible systems. What we shall concentrate on here, after a brief summary of the more important systems, are the generic principles that govern the systems' method of operation.

As their name suggests, GNSSs are primarily navigation systems – that is, they are designed to give instantaneous determinations of position and velocity anywhere on and above the surface of the Earth. Historically, they were conceived as military systems, and civilian users were a secondary consideration. With the advent of mass-market applications, this relative importance has to some extent been reversed and new systems are specifically targeting a civilian market. Whilst the vast majority of users will be making use of the 'basic' positioning functionality of the systems, in the context of this book it is important to note that there is a wide variety of different modes of operation that range in accuracy from several metres down to the centimetre or better.

Each GNSS consists of a *space segment*, a constellation of satellites in a given orbital pattern. As we shall see, the minimum requirement for any positioning mode is the ability to receive signals from at least four satellites simultaneously, and this generally dictates a constellation of twenty or more satellites at a high altitude above the Earth's surface. GPS satellites, for example, orbit at a mean altitude of 20 180 km, whilst those of Glonass are at 19 100 km. The first satellite of the Galileo system was launched in 2005 into an orbit with an altitude of 23 616 km (Dixon 2007). For each of these systems operating individually, this generally means that around 6–10 satellites are above the horizon at any one point at any one time. Of course, that does not mean that this number is always available for use, since the horizon may be obstructed by buildings or dense foliage.

The basic premise of a GNSS is the same as any surveying system: the coordinates of new points are found by making observations with respect to points of known coordinates. The only differences here are that the known points are in orbit, and they are not stationary. The determination of the coordinates of these satellites is therefore a continuous process, and is achieved by the *control segment* of the GNSS, which consists of a network of monitoring and control stations that are dedicated to the task of determining the orbital paths of the satellites and monitoring the health of their signals. It is then possible to predict the orbit of a satellite a short way into the future, and to upload this information to the satellite itself. In this way, the satellite is able to broadcast its position (referred to as the *ephemeris*, plural *ephemerides*) for the users to determine their own position in real time.

Satellites transmit information to the user's receiver by modulating it onto a set of underlying carrier waves. For the first quarter century of its existence, GPS operated with two such carrier waves, L1 and L2, with frequencies of 1 575.42 MHz (equivalent to a wavelength of 19.05 cm) and 1 227.60 MHz (equivalent to 24.45 cm), respectively. This is due to be augmented with a third frequency (L5) from 2008 onwards. Similarly, Galileo is designed to broadcast on four frequencies and Glonass on three (Leick 2004).

In addition to the ephemeris and other general information about the system, binary codes are also modulated onto the carrier waves. These can be thought of as being the equivalent of the step function shown in Figure 5.1, with each change in sign being represented by a change in phase of the carrier wave.

The combination of codes carried on each frequency is different for each system. When GPS was the only system, the only code that was available to civilian users was modulated only on one carrier wave, thus denying non-military users the higher accuracies that come from exploiting more than one frequency in the positioning process (although this may be overcome for some applications). However, with the availability of multi-frequency measurements from alternative systems, a plan for the enhancement of GPS positioning was announced in 1999 (Leick 2004).

The main point to note about the codes at this stage is that they are transmitted in a pre-determined sequence at precise times for each satellite. Thus, if the codes can be read, they can simply be thought of as time information.

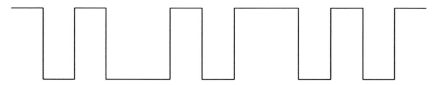

Figure 5.1 Binary code modulated onto the satellite carrier wave.

5.3 Positioning with codes

The basic method for positioning with GNSS is through the use of the codes. The procedure is for the receiver to 'read' the code, and thus obtain the time information that is transmitted by the satellite. This procedure is aided by the fact that the receiver has a copy of the code stored within it, and it knows roughly what to expect: it is therefore possible to make a reading on what is in fact a very weak signal (about half of the ambient noise level of the atmosphere).

The receiver, generating the code itself, determines the off-set in time between the code that it generates and the one received from the satellite, and from this deduces the time of travel of the signal.

By multiplying this travel time by the speed of transmission of the signal (the speed of light), the distance from the satellite to the receiver is obtained. In theory, it would be possible to determine the three-dimensional coordinates of the receiver from observations to three satellites. This follows either from considering three simultaneous equations with the three unknowns (X, Y, Z), or from a geometrical analogy in which each determination of a distance describes a sphere of a certain radius centred on the satellite, and the intersection of three spheres is at two points, one of which may be eliminated through being too far from the earth's surface.

The model described above has taken no account of the lack of synchronisation between the satellite time system and the receiver clock, however. Whilst the synchronisation between the clocks of the different satellites is achieved very precisely through the use of atomic clocks and continuous monitoring by ground control stations, the receiver has to have an independent, and much cheaper, timing system. This then leads to the problem that, since the speed of light is approximately 300 000 km/s, a one second off-set between the satellite and receiver clocks will lead to a 300 000 km error in the determination of the distance. For this reason, the distances so found are sometimes referred to as *pseudo-ranges*.

The problem is solved simply by introducing a fourth satellite, and using four distances to solve the four unknown parameters (X, Y, Z, $\Delta\tau$) where $\Delta\tau$ is the (instantaneous) off-set between the satellite system time and the receiver's clock.

The reasons for the design of the satellite constellation are now apparent: four satellites is the minimum requirement, and in fact this will virtually always be possible in the absence of any obstructions (it should be noted that the signals can travel through cloud, fog, rain, and so on, but cannot penetrate buildings, dense foliage, or similar bodies without being considerably attenuated and necessitating the use of advanced receiver design). We now see the advantage of a receiver that has been designed to receive signals from more than one system: with only one system there are typically six to ten satellites above the horizon, but the user of an in-car navigation system, for example, would

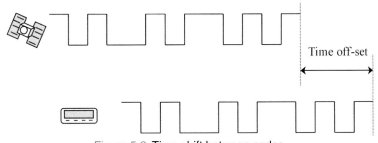

Figure 5.2 **Time shift between codes.**

often experience problems when driving down an urban street lined with tall buildings, where fewer than the required four satellites were visible. However, with access to more than one system, there are potentially sixty or more satellites available and this in turn increases the likelihood of four or more being in the visible part of the sky.

The determination of a satellite-receiver distance from code observations is subject to several sources of error. These can be summarised as follows:

- *Ephemeris and satellite clock errors*: caused by the difference between the satellite's true position and the broadcast ephemeris. Combined with the clock error, the ephemerides broadcast by the satellites are generally accurate to the sub-metre level, possibly as good as 50 cm (Creel *et al.* 2006).
- *Refraction*: caused by the difference between the actual speed of transmission and its assumed value. This error source can be sub-divided into two components: ionospheric and tropospheric refraction. The former has its origin in the upper atmosphere, and is very difficult to predict. It can cause errors of several metres in the observed range, but can be corrected by using dual frequency observations or (with slightly less success) by using a model whose coefficients are broadcast by the satellites. Without such corrections the errors can amount to several metres. Tropospheric refraction, on the other hand, is more predictable. It refers to the refraction encountered in the lowest 40 km of the atmosphere, and is dependent on temperature, pressure, and humidity. The refractive index of the dry atmosphere is very easily modelled, and its effects are therefore negligible: only the wet part of the atmosphere causes any real problems, and can be up to 40 cm for a satellite at the zenith or more for one at lower altitude (Leick 2004).
- *Multipath*. This error is caused by the satellite signal reflecting off other surfaces on its way to the antenna, and mingling with the signal that has taken a direct route: it can be minimised by good site selection or better receiver design. Its magnitude is dependent on the environment at the receiver, and can be 1–2 m.
- *Receiver noise*. This is a random error source that is related to the measuring precision of the receiver. Sub-metre precision is possible, but it depends on the quality of the receiver.

In summary it can be seen that there is a total range error to any one satellite that comprises an element due to the system itself that typically reaches an average value of around 1.1 m (Creel *et al.* 2006) and an element that is a function of the receiver and the site. The actual accuracy that will be achieved in positioning will depend on the receiver-satellite geometry. Essentially, if the distances measured from the satellites to the receiver intersect at a shallow angle (because the satellites are grouped in

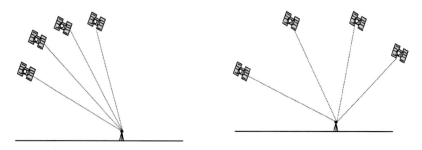

Figure 5.3 **Poor geometry (left) and strong geometry (right).**

one portion of the sky), then the accuracy of the three dimensional coordinates of the point will be worse than if the satellites were well distributed around the sky.

For satellite positioning, this concept is conveniently expressed by a numerical coefficient, known as the PDOP (Positional Dilution Of Precision). This is the ratio of the mean accuracy of the coordinated position to the accuracy of the original range obser- vations: the larger this number, the worse the geometry of the satellites. Typically, the PDOP values when using the full constellation of a system range between 2 and 5: for range errors of around 2 m to individual satellites this implies positioning accuracies of 4–10 m. If some satellites are obscured and the PDOP rises much above this range, then the accuracy would be correspondingly lower. These figures would be for the 'instantane- ous' accuracy, and averaging over time does increase the accuracy. However, most error sources (such as the ephemeris and refraction) only change very slowly and it would be necessary to average over a few days to drive the error down to even a metre.

The coordinate reference system that satellite systems use is dictated by the reference frame of the control segment monitoring and orbit determination stations. In terms of the accuracies that we are considering with simple positioning by codes, the reference systems are all compatible with each other, and can broadly be labelled as WGS 84. The actual computations of position are carried out by the receiver in geocentric Cartesian coordinates, but for display to the user these can be converted to geographic or projected coordinates.

This, then, is the basis of operation of all GNSS receivers operating as stand-alone units. It can be seen that the accuracy is mostly dependent on factors external to the receiver: any difference in price between different models is to a great extent explained by the functionality of the equipment, such as the storing of data and the use of digital

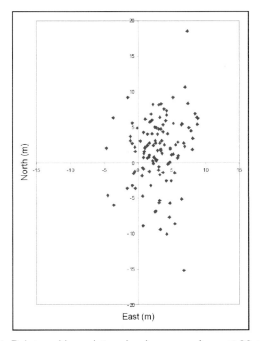

Figure 5.4 Point positions determined over one hour at 30 s intervals, with respect to the true position, using a stationary GPS receiver. Note that the mean of the one-hour data set still has a bias of around 3–4 m.

map displays, but some receivers do have lower noise or are better able to track the signals in areas of poor reception.

5.4 Differential GNSS and augmentation systems

We saw in the previous section that the accuracy of positioning with codes is limited by errors in the ephemeris and by refraction, multipath, and receiver noise. Whilst the last two are dependent on a particular site and a particular receiver, the ephemeris and refraction errors have a strong spatial correlation. That is, if we move a few kilometres away, the satellite signals are still going through more or less the same atmosphere and the refraction errors are likely to be much the same. The key to the improvement of the accuracy obtainable with GNSS is therefore the use of two or more receivers to measure *relative* as opposed to *absolute* positions.

The simplest of all augmentation systems is therefore to have one receiver set up at a point of known coordinates. In this context, 'known' implies known in the WGS 84 system, as local coordinate reference systems are entirely inappropriate for this type of computation. The information could be obtained by the mean of a long term set of absolute position observations by the receiver or, more appropriately, by linking the point to a higher order reference system through the type of observation discussed in the next section. The effect of an error in the absolute coordinates of this station will be similar to the effect of an error in the satellite ephemeris when calculating a vector from one point to another; therefore, an appropriate target for the accuracy is at least better than 1 metre.

The receiver at the known station, or the *reference receiver*, is thus able to compare the observed pseudo-range to each satellite with the value calculated from its own coordinates and the ephemeris position of the satellite. From the difference between the two, a correction for each satellite range can be determined.

The range correction so determined can then be used to correct the range observations made at another receiver, sometimes referred to as the *rover*. The efficacy of this correction will depend on the degree of correlation between the errors at the two receivers. For

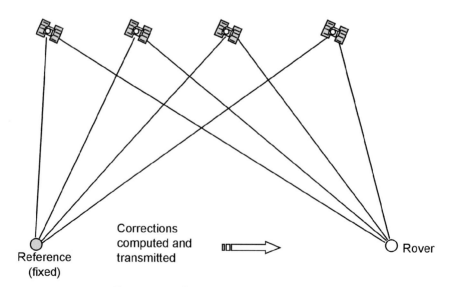

Figure 5.5 Differential GNSS arrangement.

example, an error in the satellite ephemeris will not cause an error in the range measured at the reference receiver if the direction of the error is perpendicular to the line from the satellite to the receiver. In turn, this undetected error will not cause a range error at the roving receiver if it is also perpendicular to the range, but this will be less true as the inter-station distance increases. Similarly, the degree of correlation in the effects of atmospheric refraction will decrease as the receivers become further apart.

For some applications, the corrections to the observations made at the roving station can be made after the event, once the data from the two receivers has been downloaded onto a computer. This mode of operation is suited to the determination of the coordinates of a series of static features for incorporation into a data base, for example. The alternative is to carry out the corrections in real time, in which case it is necessary to transmit the necessary information from the reference receiver to the rover. This could be done by purchasing two receivers and a radio communication link, in which case the reference receiver can be positioned wherever is appropriate for a particular project, or by a commercially provided differential GNSS service. Given the decay of the spatial correlation of errors as the distance increases, this type of set-up might be appropriate for an application such as a port authority broadcasting corrections for use within a fairly limited area. Within a range of a few tens of kilometres, this simple example of differential GNSS would be expected to give sub-metre accuracies.

Over longer distances, more sophisticated modelling of the errors is necessary. To do this, several reference stations are required, spread out over the area in which the correction service is going to operate. Instead of simply determining corrections to the observed ranges, the separate components of the errors are determined and from the different range corrections observed at different reference stations a better orbit

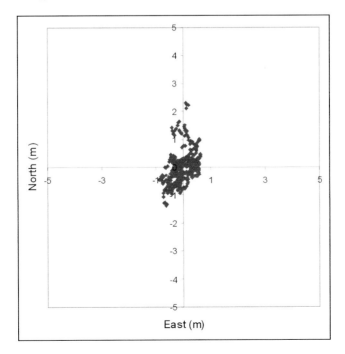

Figure 5.6 Differential positioning over one hour. Note the change in scale between this diagram and Figure 5.4.

ephemeris is determined, as well as a spatial model of the refraction effects. With the great increase in hardware and computational requirements, this is no longer an 'in-house' solution: instead, various services have been set up on an area basis to address users' needs.

It is possible to subscribe to a correction service, for example, and receive corrections via a radio link. Some such services are provided free of charge. The most significant of all correction services are what are known as *satellite based augmentation systems*. As their name implies, these operate a network of tracking stations across an area, derive correction terms, and then broadcast these to users via geostationary satellites. These correction signals are similar in format to the actual GNSS signals themselves, and so can be received by the same equipment. In this way, an improved accuracy is obtained in a fashion that is almost invisible to the user.

Notable examples of such satellite based augmentation systems are the North American WAAS (wide area augmentation system) and EGNOS (European Geostationary Navigation Overlay System). It should be noted that while improving the accuracy is an important consideration, one of the main functions of such systems is improving the integrity of the GNSS by providing rapid information about problems with particular satellites.

The end result is again a coordinate in WGS 84, but generally to sub-metre accuracy (Russell 2006). If it is required to transform to another coordinate reference system, then the methods discussed in Chapter four should again be consulted. In the context of sub-metre accuracy, we can certainly say that three parameter geocentric transformations are very unlikely to provide an accuracy that will match what has been achieved with the positioning system. Under some circumstances, the seven or ten parameter transformation discussed in sections 4.3.2 and 4.3.3 may be applicable – especially if officially endorsed – but otherwise the accuracies achievable here are now coming up against the limitations of these simple transformation models and it may be necessary to adopt the more sophisticated grid approaches discussed in section 4.4.4.

5.5   GNSS measurements using phase observations

The techniques that we have discussed so far have resulted in accuracies of a few metres down to a single metre or slightly less. Once we have made the enhancements to overcome the ephemeris and refraction problems that we discussed in the previous section, what is limiting the accuracy of positioning is the effective wave length of the codes: these are generally measured in tens or hundreds of metres, and cannot be resolved to better than a few decimetres at best. Even if they could be, multipath would still be a problem as it is a function of the chip length in the codes. The way around this is to ignore the codes (they can be removed from the signal if they are known) and go back to the underlying carrier waves. The L1 and L2 carriers on the GPS system, for example, have wavelengths of 19.03 cm and 24.42 cm, respectively. This allows the possibility of very precise observations, as measuring a fraction of a wavelength offers millimetric precision. It does this at the cost of an increased mathematical complexity of the solution, however. Whereas the code observations represent more or less unambiguous determinations of distance, the phase observations only measure a fractional part of the distance, which repeats itself every wavelength.

The basic observation is made by measuring the difference in phase between the signal received from the satellite and one generated in the receiver. The difference in phase so measured will be due to two causes. Firstly, there is a difference that is due to the fact that the receiver and the satellite are not likely to be oscillating in phase in the first place. Secondly, the fact that the signal has had to travel a certain distance from the satellite to the receiver means that the receiver is comparing its own signal with one emitted from the satellite a short time previously. In the unlikely event that the satellite were a whole number of wavelengths away from the receiver and the two signals were emitted perfectly in time with each other, then the phase difference measured by the receiver would be zero.

Hence, with reference to Figure 5.7, by converting a phase measurement into distance the receiver is actually measuring the fractional part of the distance over and above a whole number of wavelengths. Subsequent measurements in the same sequence can measure this fractional part as it goes beyond a whole wavelength, so that the initial unknown number of whole wavelengths stays the same.

The result is therefore an observation of phase that is related to the satellite-receiver range, but in a way that is complicated by the presence of an unknown whole number of wavelengths (the *integer ambiguity*) and phase differences between the satellite and receiver clocks. The key to the solution of this problem is the fact that these *bias terms* are constant, provided that the receiver continuously tracks the signal from the satellite. It is therefore possible, in time, to acquire more and more observations whilst not increasing the number of unknown parameters to be solved.

There are several different approaches that can be taken to solving the problem. Firstly, a single receiver can occupy a point for a considerable period of time (at least several hours), collecting a very large data set. In combination with a more precise ephemeris than that broadcast by the satellites, for example those disseminated via the International GNSS Service, (IGS 2007), and improved corrections to the satellite clocks, it is possible to solve all the parameters, including the three-dimensional

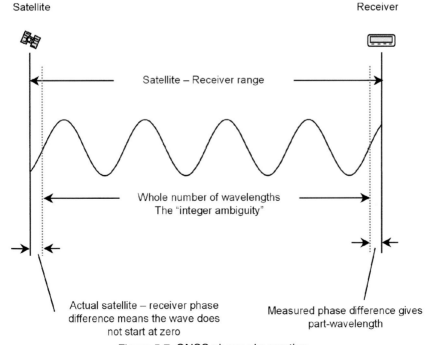

Figure 5.7 **GNSS phase observation.**

coordinates of the receiver. By definition this cannot be done in real-time, and it requires very sophisticated software, but it is capable of determining the position of the point to an accuracy of around 2 cm. This type of observation is most appropriately employed in global geodynamic studies, for example.

In this type of application, the coordinate reference frame is that defined by the ephemeris. If the broadcast ephemeris is used, then this will imply WGS 84.

Alternative, less time-intensive methods of solving the position from phase observations are based on making the observations in differential mode, using two or more receivers. Part of the advantage of this technique comes from the fact that it is no longer necessary to determine the integer number of wavelengths from the satellite to the receiver, but only the difference in the number of wavelengths from a satellite to each receiver. The real advantage of GNSS, and what has made it a tool that has all but replaced conventional control surveys, comes from exploiting the integer nature of the ambiguities, thus avoiding the need to collect data over a long period of time. Essentially the approach is to gather just sufficient data for the processing software to be able to recognise the integer values of these numbers.

Thus, for example, if the solution to the equations yielded as ambiguities the numbers:

$$451.999 \qquad 43.002 \qquad 875.998$$

then, barring coincidence, these would be recognised as 452, 43, and 876, respectively. The situation now is that the relative distances between satellites and receivers are in principle known to the precision of the phase observations, a matter of a few millimetres. This would lead to a baseline between the stations that is precise to a similar amount. Conversely, if a mistake had been made and the ambiguities incorrectly identified, then an error of several decimetres would be introduced.

Thus, the use of GNSS over short baselines is centred on the length of time required to resolve the ambiguities to the correct values *with a high degree of confidence*. Advances in modelling software have reduced the time required to a matter of minutes over baselines of several kilometres in length. The techniques involved are greatly assisted by the receivers being able to read several different wavelengths and exploiting the interference patterns between them. Generally speaking, however, these multiple frequencies are only used to acquire additional data for a quick solution and not to make any correction for ionospheric refraction: over short distances, the assumption is that this will cancel out between the two stations, which is the main factor that imposes a distance limit to this technique.

Over most distances less than a few tens of kilometres, this assumption is usually valid, and the main error source would be multipath: by definition this is site specific and does not cancel out between stations. For phase observations this can potentially lead to errors of up to 5 cm in the observation to any one satellite, and besides the obvious solution of good site selection (which is sometimes impractical) the main way around the problem is to increase the period of observation beyond the minimum required to resolve the ambiguities.

For longer baselines, the effects of ionospheric refraction are potentially more serious. For this reason, the emphasis is no longer on obtaining rapid solutions by identifying the integer ambiguities, but on correcting for the effects of refraction by exploiting multiple frequencies. This leads to much longer occupation times, usually measured at least in hours, sometimes days.

In summary, differential phase observations with GPS are capable of determining baseline vectors between receivers to a precision of around 1 cm: the length of time required for this will vary from a few minutes to a matter of days, depending on the length of the vector.

The answers so obtained will be vector components ΔX, ΔY, ΔZ. The coordinate reference frame in which they are determined will be a function of the ephemeris used; the coordinate reference frame in which the coordinates of new points are obtained will also depend upon how these vectors are combined with fixed points of known coordinates. The coordinate reference frame of newly surveyed points can only really make sense if the reference frame of the ephemeris and the reference frame of the fixed coordinates are equivalent or parallel at the accuracy required. Such a condition would not be satisfied, for example, when measuring vectors in WGS 84 with respect to a national mapping system referenced to a datum defined using historical techniques; it would hold, however, for vectors measured in WGS 84 with respect to control points in ETRS89, since the two systems are offset but parallel. The coordinates of the new points would be in ETRS89.

In other applications, it is sometimes convenient to start with a reference station whose WGS 84 coordinates are determined not from a higher order survey but from the average of several hours of code observations of the type described in section 5.3 above. In this case, the vectors are accurate but the coordinates established are 'floating' with respect to true WGS 84 by an offset of a metre or more, and therefore a high accuracy grid based transformation of the type discussed in section 4.4.4 cannot be used unless a preliminary transformation is carried out to correct the original absolute coordinates of the reference station.

Differential phase observations could be made by a surveyor using two or more receivers and processing the results on a commercial software package. These are generally designed to process vectors up to a few tens of kilometres in length. In other situations, for example when determining the coordinates of an initial geodetic control point from which other vectors will establish a national coordinate framework, it may be necessary to carry out observations over hundreds or thousands of kilometres, which not only requires more sophisticated software but presents logistical problems when setting up base stations. In these situations, recourse may be made to web-based services such as Auto-GIPSY in which data may be submitted via the internet and coordinate values returned (UNAVCO 2007). In these situations, *relative* positioning is virtually indistinguishable from *absolute* positioning as far as the user is concerned. Around ten hours or more of data are generally required to achieve centimetric accuracy (Ghoddusi-Fard and Dare 2006).

A slightly lower level of accuracy can be achieved more quickly over shorter distances through the use of *kinematic* techniques. The basic premise of these is that once the initial determination of the integer ambiguities has been made, these values will stay the same even when the roving receiver moves to a new location, provided that the receiver continuously tracks the signal from the satellite. As with differential GNSS using code observations, it is possible to operate kinematic phase GNSS in a real-time mode. This is usually referred to simply as real-time kinematic, or RTK. The data link can be provided by a dedicated radio link or, in a parallel development to those seen with code positioning, can be delivered by mobile phone as a commercial service. The latter works through a technique known as the *virtual reference station* in which a network of real reference stations models the corrections that would be obtained at a virtual station in the vicinity of the user. Examples of this type of service are those based on the Ordnance Survey of Great Britain's OSNet® system (Ordnance Survey 2007).

In principle, the accuracy of kinematic GNSS is similar to that of conventional phase observations in the static mode. It will always be slightly lower, however, as there is no averaging over time, which means that the effect of multipath goes uncorrected. Typically, at any one point there will be an error of around 2 to 3 cm.

The procedure for kinematic GNSS is first to initialise the roving receiver by acquiring enough data to resolve the integer ambiguities. This can be done whilst the receiver is stationary or, with 'on the fly' (OTF) techniques, data may be collected whilst the receiver is in motion. For OTF, the position of the receiver during the initialisation period can be deduced after the event, but not in real time.

Even if an individual signal is interrupted, it is possible to proceed provided that at least four satellites are tracked continuously. This is because four satellites are the minimum requirement for positioning, and if the position is known, when new satellites are observed, or new ones re-acquired, the resolution of the ambiguities is instantaneous. A complete blockage of all signals, such as would be caused by passing under a bridge, cannot be supported, however. Under these circumstances, it is necessary to go through the initialisation cycle once again: repeated interruptions such as would occur in a cluttered urban environment would cause problems.

5.6 Coordinate reference system considerations

GNSS systems such as GPS operate in their own global coordinate reference system. In the case of GPS this is WGS 84. The ubiquitous nature of GPS has led to WGS 84 becoming the coordinate reference system of choice for many applications, but it may not always be appropriate. What are the options if we wish to use something different? We need to transform the GPS coordinates into our preferred coordinate reference system. The methods to do this are discussed in Chapter 4.

There may be an option of performing the transformation using the GPS receiver software. This has the advantage of displaying coordinates in your desired system, but some care is needed. Most receivers store positions in WGS 84 terms and transform to local values only for display purposes. If you download the positions, these may still be in WGS 84, even when the receiver display has been set to show the local system. And the download output may not be labelled to indicate which coordinate reference system has been used.

Some receivers have the ability to display coordinates after applying one of a number of in-built transformations. Most use the Molodensky method for the transformation. As discussed in section 4.4.2, this is equivalent to a three-parameter geocentric transformation. Such transformations typically have an accuracy of about 5 to 10 m; that is, if the GPS coordinates were considered error-free (keep on dreaming!) then the result of applying the transformation will give local coordinates within about 5 to 10 m. Where they are available, officially sanctioned transformations, particularly those using bilinear interpolation of a gridded data set, are usually accurate to fractions of a metre. And for the reasons discussed in section 4.8, by preference they should be used.

Figure 5.8 shows the practical application of different transformations applied in an in-car navigation setting. In this figure, the track of a car has been monitored using differential GPS and plotted on a map based on a national coordinate reference system. The Molodensky transformation (section 4.4.2) that uses the three-parameter geocentric translations is shown as the red line in the diagram, and as expected is 10 m or more from the correct path. This is improved by using a seven-parameter transformation with parameters that are appropriate to the country as a whole: the blue line

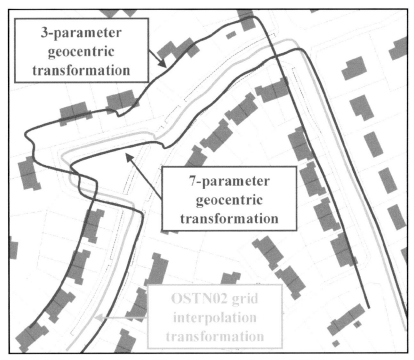

Figure 5.8 The application of different transformations to GPS data. Transformations applied use best national parameters for the given transformation method.

on the diagram is shown to follow the road correctly, and its errors here of a couple of metres would be appropriate for many in-car navigation applications. However, the official OSTN02 transformation, using bilinear interpolation of gridded data (section 4.4.4) can be seen in this example to show which side of the road the car was driving on, and therefore could be used in navigation systems to instruct the driver on lane changes and so on.

If the GPS receiver is unable to use the official transformation and the user does not wish to use the internal transformation there are two options. One is to have the receiver work in WGS 84 coordinates and transform the coordinates later using an external application, or in charting applications using the recommended shifts in latitude and longitude that are printed on the chart. (The receiver manufacturer might also be badgered to include the official transformation in their firmware!) The second is to derive a personal Molodensky transformation appropriate for the survey area and use this in the receiver. Case Study 6.2 shows an example of how the transformation parameters may be 'tuned' for the local area. A similar approach can sometimes be taken with high precision applications such as setting out with RTK. If a single control point is available on a construction site then this can be surveyed with the roving GPS receiver, the control point coordinates keyed in, and the parameters of a Molodensky transformation or its equivalent instantaneously derived. These parameters would be applicable in the immediate vicinity of the control point, but it should be remembered that a three-parameter transformation is ignoring any rotation of the local coordinate reference system with respect to WGS 84. With rotations of 10" or so, and similar changes in the geoid-ellipsoid separation, the centimetric accuracy of the survey is lost within 200 m.

6

CASE STUDIES

6.1 Transformation of GPS data into a local coordinate reference system

This case study will examine the options that are available for transforming position data acquired with differential phase GPS observations into a national coordinate reference system. Twelve new stations A to M have been related to four national triangulation pillars. Figure 6.1 shows the configuration of the points observed, although the vectors are not shown to avoid cluttering the diagram. The network adjustment of the new observations indicates that the quality of the vectors is on average around 1.5 cm in plan and 2 cm in height.

The first point to note is that this example of an observational network happens to be in Great Britain. Therefore, in principle, only one transformation is acceptable: the *official* grid-based OSTN02 transformation of the type discussed in section 4.4.4,

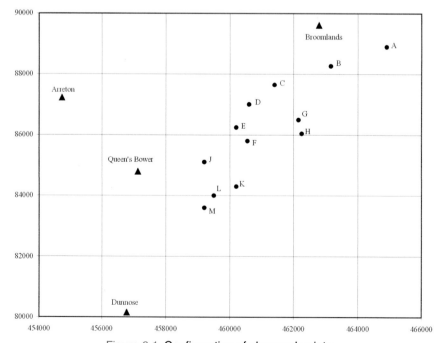

Figure 6.1 Configuration of observed points.

together with the OSGM02 height transformation (section 4.7.2). However, since there are many parts of the world that do not have such official models available, we shall examine the cases that result from adopting alternative techniques. We shall look at the 3-, 7- and 10-parameter geocentric methods discussed in section 4.3 and the Similarity transformation method as implemented in some GPS processing software and discussed in section 4.5.7.

A necessary requirement for deriving transformations to be applied to accurate observational data is a set of points that have known coordinates in the local system. This has been achieved by incorporating several Ordnance Survey triangulation pillars in the survey, as shown in Figure 6.1 by the named points. A reasonable spread of control points has been achieved: it was not possible to surround the area completely as the south east section of the map is in the sea. Several bench marks have also been included in the survey.

The local coordinates are given in the form of Eastings and Northings in terms of the British National Grid, and heights in the Newlyn vertical coordinate reference system (ODN). They therefore constitute a compound coordinate reference system. If a geoid were available, it would be advisable to first transform to a complete three-dimensional ellipsoidal system. However, we shall assume for this exercise that no such model can be obtained.

Three-parameter geocentric transformation

The simplest transformation is the three-parameter geocentric one described in section 4.3.2. We start by converting the national E,N coordinates of the triangulation pillars into geographic ones using the appropriate map projection formulae and parameter values. We then assign the gravity-related heights in ODN as 'stand-ins' for the ellipsoidal heights relative to the national ellipsoid, and have therefore obtained a set of 'dummy' ellipsoidal coordinates in the national geodetic coordinate reference system. These are then converted to XYZ geocentric coordinates in the national system, and we are then ready to determine the transformation parameters.

Comparing the geocentric Cartesian coordinates in WGS 84 at the triangulation pillars from our new network adjustment with the values in the national system, and applying the least squares techniques described in Appendix E, we obtain:

Table 6.1 **Three-parameter geocentric transformation derived at four points by least squares.**

ΔX (m)	ΔY (m)	ΔZ (m)	**Standard deviation (m)**
370.153	−109.582	435.572	0.067

The values obtained here differ from any other set published for this transformation for several reasons. In the first place, published values are generally for the coordinate reference system as a whole, whereas this is just for a local sub-set: it is therefore tailored better to the local distortions. Secondly, we made that rather sweeping assumption about the gravity-related heights being the same as ellipsoidal, and so a component of this shift will be dealing with the average geoid-ellipsoid separation across the area. And thirdly, our WGS 84 coordinates were not obtained by tying the network to

Table 6.2 Residuals from the three-parameter geocentric transformation.

	Residuals (m)		
	Latitude	Longitude	Height
Arreton	−0.050	0.030	−0.003
Broomlands	−0.048	−0.067	−0.097
Dunnose	0.106	0.027	0.091
Queen's Bower	−0.008	0.011	0.010

a continuously operating ITRF-related reference system but by taking the average of a few hours of observations at one point. The absolute position in WGS 84 is therefore wrong by about half a metre, but the relative positioning is accurate to the 1–2 cm previously quoted.

Because three coordinates – two horizontal and height – were known at each triangulation pillar in both the national coordinate reference system and observed as WGS 84, it would have been possible to derive a 3-parameter geocentric transformation at any one of the stations. Each would have given slightly different transformation parameter values, both from each other and from those in Table 6.1 derived by a least squares fit to all four pillars. The value of a least squares fit comes from examining the residuals – the difference in coordinates resulting after application of the derived transformation with their actual values. When deriving a geocentric transformation, most applications will give these in the geocentric XYZ domain. We have transposed them into horizontal and vertical to examine trends in these domains.

We can see in Table 6.2 that overall the residuals are slightly greater than we believe the accuracy of our survey to be. We therefore conclude that the transformation method is inappropriate for the level of accuracy we require over this area.

More noticeable from Table 6.2 is that the heights have much larger residual values than the 2 cm suggested in the quality analysis of our new network adjustment. There is a trend from north-east (largest negative residual) to south-west (largest positive residual). Although with such limited data it is difficult to draw conclusions, it is quite possible that there is a geoid slope across the area. It suggests that our assumption that national gravity-related height could be taken as an ellipsoidal height was not valid.

To illustrate the value in the examination of residuals better, we have introduced a 25 cm error in the national system position of Broomlands. Now our least squares analysis produces different transformation parameter values and residuals as shown in Tables 6.3 and 6.4, respectively.

Table 6.3 Three-parameter geocentric transformation derived using erroneous coordinate set.

ΔX (m)	ΔY (m)	ΔZ (m)	Standard deviation (m)
370.151	−109.650	435.572	0.121

Table 6.4 Residuals from the three-parameter geocentric transformation using erroneous coordinate set.

	Residuals (m)		
	Latitude	Longitude	Height
Arreton	−0.049	0.097	−0.003
Broomlands (with 25 cm error)	−0.051	−0.270	−0.097
Dunnose	0.107	0.094	0.091
Queen's Bower	−0.007	0.078	0.010

The first point to notice here is that the standard deviation of the least squares fit is significantly higher than we believe both our new survey and the national network accuracy to be. This should set alarm bells ringing. When we look at the residuals, they are in general larger than we would anticipate and that for the longitude of Broomlands is significantly higher than we would expect and should be treated with suspicion. Deriving another transformation with Broomlands omitted would be appropriate.

Seven-parameter geocentric transformation

The next transformation method that is tried is the 7-parameter transformation of section 4.3.3. The transformation parameter values and residuals that result from this are shown in Tables 6.5 and 6.6.

The standard deviation and residuals of this 7-parameter transformation are mostly within the accuracy of the GPS survey. The method appears to have produced a better solution than did the 3-parameter geocentric method (see Tables 6.1 and 6.2). In particular the perceived geoid slope seems to have been incorporated into the solution. However,

Table 6.5 Coordinate frame geocentric transformation derived at four points by least squares.

ΔX (m)	ΔY (m)	ΔZ (m)	α_X (sec)	α_Y (sec)	α_Z (sec)	μ (ppm)	**Standard deviation (m)**
507.523	−157.534	446.091	0.879	3.223	−1.312	−15.11	0.019

Table 6.6 Residuals from the seven-parameter coordinate frame transformation derived at four points.

	Residuals (m)		
	Latitude	Longitude	Height
Arreton	−0.012	−0.014	0.001
Broomlands	0.009	0.018	−0.001
Dunnose	0.030	−0.002	0.000
Queen's Bower	−0.012	−0.001	−0.008

Table 6.7 Residuals from the coordinate frame seven-parameter transformation using erroneous coordinate set.

	Residuals (m)		
	Latitude	Longitude	Height
Arreton	0.003	0.030	0.002
Broomlands (with 25 cm error)	0.009	−0.049	0.000
Dunnose	−0.039	−0.017	0.003
Queen's Bower	−0.011	−0.050	−0.009

the three translation parameters ΔX, ΔY and ΔZ now have very different values to those we obtained from the 3-parameter solution (Table 6.1). This does not reflect reality, since they change to compensate the rotations about the centre of the ellipsoid.

Before proceeding further, however, it is worth examining the extent to which an error in the control points would be noticeable by an examination of the residuals. To do this, the error was reintroduced into the longitude coordinate of Broomlands, increasing it by 25 cm. The residuals that result from this transformation are shown in Table 6.7.

It is clear that the error has been absorbed into the transformation and that there is not sufficient redundancy to isolate the problem to the coordinates of Broomlands.

Ten-parameter geocentric transformation
The next transformation to be tried is the 10-parameter Molodensky-Badekas transformation of section 4.3.4. This gives the parameter values and residuals of Tables 6.8 and 6.9.

Note the characteristics of these results:

- The three translations are almost the same as those from the 3-parameter method given in Table 6.1. The 10-parameter method therefore models the offset of the ellipsoid centres significantly better than does the 7-parameter method.
- The three rotations and the scale change are identical to those from the 7-parameter method given in Table 6.5. So too is the standard deviation of the transformation. The 10-parameter method is therefore modelling the orientation and scale changes as well as the geoid slope in the same way as does the 7-parameter method.
- The 10-parameter method's residuals (Table 6.9) are identical to those from the 7-parameter method given in Table 6.5. When applied, the 10-parameter transformation gives identical results to the 7-parameter method.

Table 6.8 Molodensky-Badekas ten-parameter transformation derived at four points.

ΔX (m)	ΔY (m)	ΔZ (m)	α_X (sec)	α_Y (sec)	α_Z (sec)	μ (ppm)	Standard deviation (m)
370.150	−109.580	435.570	0.879	3.223	−1.312	−15.11	0.019
$X_O = 4049730.54$ m			$Y_O = -83504.08$ m			$Z_O = 4909751.63$ m	

Table 6.9 **Residuals from the Molodensky-Badekas ten-parameter transformation.**

	Residuals (m)		
	Latitude	Longitude	Height
Arreton	−0.012	−0.014	0.001
Broomlands	0.009	0.018	−0.001
Dunnose	0.030	−0.002	0.000
Queen's Bower	−0.012	−0.001	−0.008

In summary, the 10-parameter method combines the modelling characteristics of the 3- and 7-parameter methods.

Similarity transformation method

As a final example, we examine the two-dimensional transformation of the WGS 84 coordinates into the national projected coordinate reference system as described with reference to GPS projects in section 4.5.7. In this case, the ellipsoidal WGS 84 coordinates are first converted to a temporary Transverse Mercator system with an origin at the centre of the survey area, then the four parameters of the Similarity transformation method are determined as:

Table 6.10 **Similarity transformation derived at four points by least squares.**

Geometric form	X_{TO}	Y_{TO}	α	μ	Standard deviation (m)
	459787.826	84878.038	−0°39'11.179"	0.99966574	0.021

Parametric form	a_0	b_0	a	b	
	459787.826	84878.038	0.999600791	−0.0113948	0.021

Table 6.11 **Residuals from the Similarity transformation method derived at four points.**

	Residuals (m)	
	East	North
Arreton	0.025	−0.004
Broomlands	−0.014	0.013
Dunnose	−0.020	−0.016
Queen's Bower	0.007	0.007

On the face of it, the standard deviation and residuals given in Tables 6.10 and 6.11 look perfectly acceptable. The problem with applying this method in these circumstances is that the scale factor of the map projection dominates the solution and masks any ability to detect problems with the coordinates. This can be seen through applying the method to the erroneous coordinate set as seen in Tables 6.12 and 6.13.

Table 6.12 Similarity transformation derived using erroneous coordinate set.

Geometric form	X_{TO}	Y_{TO}	α	μ	Standard deviation (m)
	459787.916	84878.004	$-0°39'08.438''$	0.99968144	0.058
Parametric form	a_0	b_0	a	b	Standard deviation (m)
	459787.916	84878.004	0.999616498	-0.011395	0.058

Table 6.13 Residuals from the Similarity transformation method using erroneous coordinate set.

	Residuals (m)	
	East	North
Arreton	-0.017	-0.073
Broomlands (with 25 cm error)	0.056	0.013
Dunnose	0.000	0.052
Queen's Bower	-0.040	0.008

The 25 cm coordinate error cannot be detected from the adjustment statistics given in Tables 6.12 and 6.13.

This method distorts the scale of the GPS data into the local system. It has application when the definition of the local system is unknown, but applying it to transform GPS coordinates into an identified projected coordinate reference system is a seriously bad idea. Things could have been worse than this example suggests. The study area happens to be near the central meridian of the national grid, so the grid convergence and rate of change of scale are low. At another location, where the projection distortions are more noticeable, the method would average these distortions across the survey area.

6.2 Creation of a three-parameter geocentric transformation from an official national transformation

This case study will examine the suitability of transformations built into GPS receivers and demonstrate the derivation of Molodensky method parameters from a grid interpolation transformation. We are working on a project covering a few square kilometres in the north of Scotland. We have a GPS receiver that is programmed with a Molodensky transformation (section 4.4.2) allowing coordinates to be shown either in WGS 84 terms or transformed to one of several local coordinate reference systems, including one for Great Britain. The receiver also allows input of a 'user datum'. By

Table 6.14 Molodensky transformation parameter values in a GPS receiver.

	Great Britain (mean)
From Source CRS	OSGB 1936
To Target CRS	WGS 84
Parameter	Parameter value
Δa	+573 m
Δf	+0.1196002
ΔX	+375 m
ΔY	−111 m
ΔZ	+431 m

reference to the receiver user manual or to (NGA 1990; 2001), we believe that the Molodensky parameter values built into the receiver are as in Table 6.14.

We now evaluate this transformation at a test point by setting a waypoint in the receiver with the receiver set to reference WGS 84, switching the setting to use the inbuilt Great Britain system, and noting the output. We compare this with the official transformation, available at the national mapping agency's website as either an online calculation or as a download. The comparison is given in Table 6.15.

We decide that the difference of 10 m is unacceptable for this project.

Using the official transformation, we now calculate geocentric coordinates for several points surrounding our area of interest in both WGS 84 and local terms.

We then calculate the geocentric translations $\Delta X, \Delta Y, \Delta Z$ at each point (Table 6.16) noting that each of the differences have similar values at each point and we accept the mean values for each translation. These are input into our GPS receiver and, after using the waypoint to check they are correct, used during the project. Throughout, care is needed to ensure that signs are correct. Over the project area this transformation will have an accuracy of about 0.1 m.

Table 6.15 Evaluation of GPS receiver and official transformations.

	Waypoint in WGS 84			**Transformed to Local CRS**		
				British National Grid		ODN
	Latitude (deg)	**Longitude (deg)**	**Ellipsoidal height (m)**	**Easting (m)**	**Northing (m)**	**Height (m)**
Official transformation	57.5	−1.8	100	412085.8	845551.4	50.9
GPS receiver	57.5	−1.8	100	412084.1	845541.5	51.3
Difference (error in GPS receiver transformation)				−1.8	−9.9	0.4

Table 6.16 **Geocentric coordinates from official transformation.**

WGS 84			Local			Local to WGS 84		
X	Y	Z	X	Y	Z	ΔX	ΔY	ΔZ
3427136.5	−109767.9	5360067.0	3426752.9	−109654.8	5359641.1	383.6	−113.1	426.0
3427238.3	−106768.5	5360062.5	3426854.7	−106655.3	5359636.5	383.6	−113.2	425.9
3432299.7	−106945.2	5356841.0	3431916.3	−106832.1	5356415.0	383.5	−113.1	426.0
3432198.0	−109944.6	5356845.6	3431814.4	−109831.5	5356419.6	383.6	−113.0	426.0
					Mean	383.6	−113.1	426.0

6.3   Designing a map projection

This case study will explore the procedures that are followed in designing a map projection for a particular purpose. In this case, the example selected is to design a projection for use over the whole of Australia.

The problem can be stated in the following form:

Base maps of Australia are produced on a different projection or series of projections for each individual state. This is due to the fact that the original survey work had to be computed on projections that kept the scale factor distortion to a very low level. A company has acquired the data and wishes to put it all onto one projection, in order to avoid having discontinuities in the data. Area is not important, as this will be an attribute on the database, but features should 'look right'; that is, there should not be any great distortion of shape. Design a suitable projection.

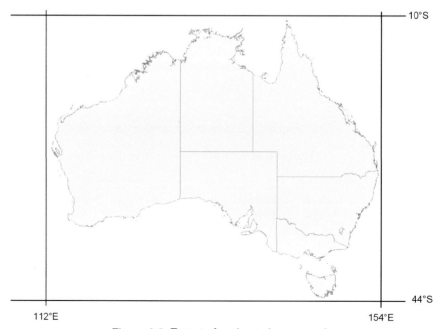

Figure 6.2 **Extent of region to be mapped.**

The approach adopted is as follows.

If features are to 'look right', then a conformal projection is called for, as this minimises the distortion of shape.

Then, in selecting a suitable developable surface, it is noted that the extent of the region to be mapped ranges in latitude from 44°S to 10°S, and in longitude from 112°E to 154°E. The area could therefore be described as a mid-latitude one with a broad spread in longitude, which leads to the suggestion of a conic projection. As it has already been decided that a conformal projection is required, the Lambert Conformal Conic projection is selected.

In selecting the required parameters, it was stated in section 3.5.4 that the scale factor distortion is minimised when the standard parallels are selected as being 1/6 of the range of latitude in from the extremes of the projection. Since the range of latitude is 34° (44–10°S), this suggests 16°S and 38°S as suitable standard parallels. A plot of the scale factor as a function of latitude is shown in Figure 6.3, on which the latitudes of the main population centres have been marked.

With the parameters selected, the minimum value of scale factor is 0.98 and the maximum is 1.03, which means a maximum distance distortion of 3%. Any areas measured from the map will be distorted by the square of this value, or 6% at the most. These levels of distortion are fine for many low-accuracy applications in which the convenience of having a single coordinate reference system across a region is paramount, but would certainly present problems for precise computations.

The central meridian, or the longitude of origin, is selected as the mean of the extremes of longitude. This is 133°E. The maximum longitude difference from the central meridian is then 21°, which gives an indication of the convergence. From equation 3.31 of section 3.5.1, and using the average standard parallel value of 27°S for α, the maximum value of convergence is seen to be around 10°. This is depicted in Figure 6.4.

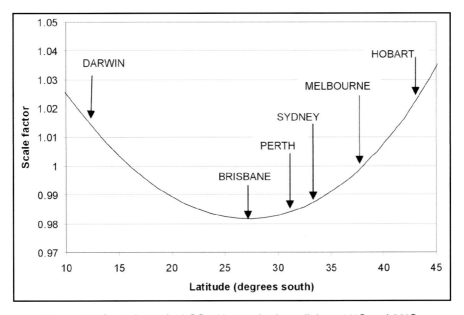

Figure 6.3 Scale factor for LCC with standard parallels at 16°S and 38°S.

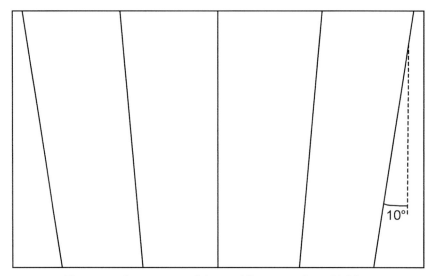

Figure 6.4 **Appearance of the meridians on selected LCC.**

The final parameter values to be selected are the latitude of origin and the values of the false grid coordinates. The choice of origin is not a critical decision, as it in no way affects the shape of the projection: for convenience φ_0 is then selected as 45°S, which gives all points in the region positive northings. The false northing can then be selected as zero.

Eastings are calculated as positive and negative quantities from the central meridian. A false easting of +2500 000 m would be sufficient to make all values positive.

This then is the map projection that has resulted purely from looking at the region to be mapped and considering the requirements of the application. It is instructive to compare this with a projection that has been 'officially' adopted for Australia, the *Geoscience Australia Standard National Scale Lambert Projection* (OGP 2007a). There is a slight difference in the longitude of origin: 134°E has been selected, rather than 133°E. More significantly for the scale factor, the standard parallels chosen are 18°S and 36°S: a comparison of the two scale factors is shown in Figure 6.5.

It can be seen that on the Geoscience Australia map projection the scale factor is closer to unity across the majority of the country and certainly over the latitudes of Brisbane, Perth, and Sydney. At Melbourne, the distortion is higher, though still less than 1%. This improvement across the main land mass has been brought about at the expense of increasing the distortion at the latitude extremes to around 3.5%.

The conclusion to be drawn from this is that in designing a projection we have followed a logical path and selected something that is fit for purpose as we see it; however, somebody else going through the same procedures may make slightly different judgements and assign different levels of importance to aspects of the projection, and therefore come up with a different solution.

6.4 Calculations using map grid coordinates

This case study will examine aspects of calculations within the projected coordinate reference systems that were discussed in section 3.8. We will examine the calculation

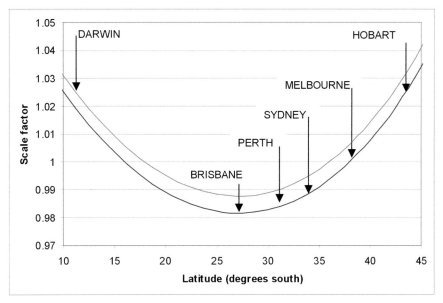

Figure 6.5 Scale factor for designed map projection (blue) compared with that for Geoscience Australia (pink).

of distance and area on a conformal map projection, using UTM as the example, and then look at issues associated with the transfer of cell-based grids from one projected coordinate reference system to another.

We have received from an external organisation a dataset based on a regular grid and wish to merge it with our own data. Unfortunately, the coordinate reference system they are using – WGS 84 and UTM zone 33N – differs from that we are using for our work, WGS 84 / UTM zone 34N.

The grid definition for the external organisation's grid has been provided. Each cell is 25 m × 50 m. The study area is 30 × 20 km and there are therefore 1200 × 400 cells over the whole area. To better distinguish between row and column numbers, 5000 is added to row numbers. The grid is geo-referenced to the UTM projection and WGS 84 datum; the projected grid coordinates of the centre of the first cell are given and the grid is orientated west–east. The arrangement is depicted in Figure 6.6.

The coordinates of the centres and corners of all cells may be readily calculated from this definition within the WGS 84 / UTM zone 33 coordinate reference system upon which the grid is defined.

Distance calculations

The first thing to note is that this grid has been constructed on a map projection. It therefore inherits all of the properties of that map projection. The sides of the grid are 30 × 20 km in grid terms only. The true distance of, for example, the northern side AB can be calculated through formulae given in advanced survey texts, such as Bomford (1980), as 29 991.31 m.

We saw in section 3.8 that the scale factor for a line may be calculated by using the mean of the point scale factor at each end of the line, or through using Simpson's

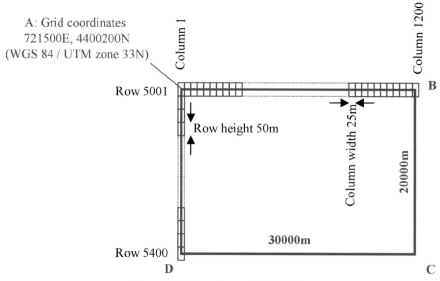

Figure 6.6 **Sampling grid definition.**

rule (equation 3.41) by additionally incorporating the point scale factor at the centre of the line.

For the example in this case study, the scale factor on the original projected CRS for the northern boundary is:

Table 6.17 **Line scale factor on initial CRS for study area side AB.**

Cell	Column	Row	Location	Point scale factor
Top left (A)	1	5001	Centre	1.00020407
Top mid	600	5001	Centre	1.00028866
Top right (B)	1200	5001	Centre	1.00037880
				Line scale factor
(a) Line AB scale factor using average of end points only				1.00029144
(b) Line AB scale factor using Simpson's Rule				1.00028959

Then through equation 3.38, we can calculate the true length of the line AB as:

a) Using side AB line scale factor derived from end points only: 29 991.26 m, which is about 2 parts per million in error when compared with that derived rigorously as given above.
b) Using side AB line scale factor derived through Simpson's rule: 29 991.31 m, which compares well with that derived rigorously.

Area calculations

The distortions in length in a conformal projection also affect area measurements. For example, the original study projection area is 600 km² measured on the map. But its 'true' area calculated using formulae given in advanced survey texts after projecting the grid boundary onto the surface of the ellipsoid is 599.652633 km². Had the origin of the grid been at a different point or had the grid been orientated differently, the same 600 km² on the map would have equated to a different true area.

We can estimate the true area of the grid. For the general case of an irregular polygon, the weighted mean of the point scale factor at each vertex plus 4 times the point scale factor at the midpoint of each side will usually give an estimate of the true area accurate to several parts per million (ppm). But *for a rectangular area in low and mid latitudes*, the average of the line scale factor for each of two particular sides gives an approximation that is better than 1 ppm for a rectangle orientated at 45° to the map grid and significantly better than this when the rectangle is orientated parallel with or perpendicular to the map grid, as is the case in this example. The two sides in question are those that traverse the greatest change in point scale factor; for a Lambert Conic projection in which scale factor change is predominately north-south this will be the east and west sides of the rectangle; for a Transverse Mercator projection (as in this example) the average of the line scale factor for the north and south sides should be used.

For the example in this case study, the point scale factor on the WGS 84 / UTM zone 33N projected CRS used by the external agency is:

Table 6.18 Point scale factor for bounding lines of grid.

Cell	Column	Row	Location	Point scale factor
Top left (A)	1	5001	Centre	1.00020407
Top mid	600	5001	Centre	1.00028866
Top right (B)	1200	5001	Centre	1.00037880
Right mid	1200	5201	Centre	1.00037882
Bottom right (C)	1200	5400	Centre	1.00037883
Bottom mid	600	5400	Centre	1.00028869
Bottom left (D)	1	5400	Centre	1.00020409
Left mid	1	5201	Centre	1.00020408

From these point scale factors, the average line scale factor may then be calculated:

Table 6.19 Ellipsoidal area for study area ABCD.

Method of derivation	Line scale factor	Ellipsoidal Area (km²)
(a) Average point scale factor at four corner points only	1.00029145	599.650416
(b) Weighted average at four corner points and four midpoints	1.00029034	599.651744
Northern side AB line scale factor using Simpson's Rule	1.00028959	
Southern side CD line scale factor using Simpson's Rule	1.00028961	
(c) Average of northern and southern Simpson's Rule values	1.00028960	599.652632
Rigorous computation		599.652633

Then using equation 3.39 to calculate the areas given in Table 6.19:

(a) Applying the average point scale factor of only the four corner points to the grid area of 600 km^2 gives an ellipsoidal area which is about 4 ppm in error when compared to the rigorously computed value.

(b) Applying the general polygon weighted average of corner and boundary midpoints gives an ellipsoidal area which is about 2 ppm in error. This has worked reasonably well for our study area, which is orientated east-west, but in a more irregular example it would not be so good.

(c) Application of Simpson's rule to find a more representative value for the line scale factor for the northern and southern sides of the rectangle and meaning these line scale factors gives the ellipsoidal area to an accuracy of one part in 10^9.

If the area approximation is not sufficiently accurate, then the options are either to use complex calculations on the surface of the ellipsoid or, for area measurements, to create a grid on an equal area projection. For statistical sampling, this latter approach is recommended.

We now wish to change this external agency grid to the projected CRS we are using, the neighbouring UTM zone with a central meridian 6 degrees farther east than the other agency. (They and we use the same geodetic datum, but had they used a different datum to ourselves, then the conversions below would have to have been supplemented with a datum transformation.) We therefore convert the grid cell coordinates from the external agency CRS to our CRS using the indirect coordinate conversion technique described in section 4.5.8. Coordinates in both systems for selected locations are given in Table 6.20.

Using the converted coordinates of the corners ABCD and plane geometry, we can calculate distance and area in the WGS 84 / UTM zone 34N projected coordinate reference system. We find that the sides AB and CD are no longer 30 000 m. And the area on the map grid is no longer the 600 km^2 it was on in the zone 33 projection but in the zone 34 map grid is now 600.319 km^2. Had the origin of the grid been at a different point or had the grid been orientated differently in the zone 33 CRS, although the area on the zone 33 map would have remained the same 600 km^2, the area on the zone 34 map would have not been the 600.319 km^2 we have just calculated.

Shape

Using the new coordinates from Table 6.20, we can calculate within our WGS 84 / UTM zone 34N projected CRS the vector distances and internal angles at the four corners of the grid, as shown in Figure 6.7.

We see that the size *and shape* of the study area grid have apparently changed. Opposite side lengths are no longer equal or parallel, and the corner angles are no longer right angles. The whole area has been warped. Further examination shows that the warping is not consistent across the area; individual cells are warped differently, as depicted in Figure 6.8.

What is happening here? Why is the angle in the top left corner of cell 1,5001 different to that for the whole grid?

Table 6.20 Coordinates of grid cells in original and new projected CRSs.

Cell	Column	Row	Location	WGS 84 / UTM zone 33N		WGS 84 / UTM zone 34N	
				Easting	Northing	Easting	Northing
Top left (A)	1	5001	Centre	721500.00	4400200.00	207231.85	4402587.02
Top right (B)	1200	5001	Centre	751500.00	4400200.00	237172.46	4400579.33
Bottom right (C)	1200	5400	Centre	751500.00	4380200.00	235836.91	4380622.44
Bottom left (D)	1	5400	Centre	721500.00	4380200.00	205895.52	4382622.57
Top left	1	5001	Top left corner	721487.50	4400225.00	207221.05	4402612.81
Top left	1	5001	Top right corner	721512.50	4400225.00	207246.00	4402611.14
Top left	1	5001	Bottom right corner	721512.50	4400175.00	207242.66	4402561.23
Top left	1	5001	Bottom left corner	721487.50	4400175.00	207217.70	4402562.90
Bottom right	1200	5400	Top left corner	751487.50	4380225.00	235826.10	4380648.22
Bottom right	1200	5400	Top right corner	751512.50	4380225.00	235851.05	4380646.56
Bottom right	1200	5400	Bottom right corner	751512.50	4380175.00	235847.71	4380596.66
Bottom right	1200	5400	Bottom left corner	751487.50	4380175.00	235822.77	4380598.33

The vectors shown in Figures 6.7 and 6.8 do not exactly represent the track of the grid boundary projected into the new CRS. The warping is not linear, and produces a continuous change depicted in exaggerated form by the solid line in Figure 6.9.

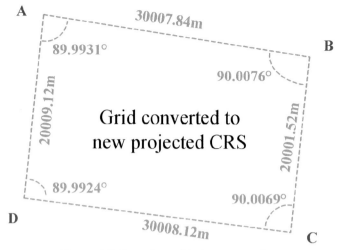

Figure 6.7 Sampling grid after conversion.

The warping is caused by the continuous change in scale and convergence across each of the two conformal map projections. Through the two projections, we have a double application of the difference between the projection line and actual line shown in Figure 3.46. We noted in section 3.2.4 that conformal projections preserve shape only at a point; when distance is introduced the property is not maintained. The warping in any individual cell a few tens of metres across is small and usually insignificant. Over a large area, the warping exceeds the cell size and, as we see here, becomes significant.

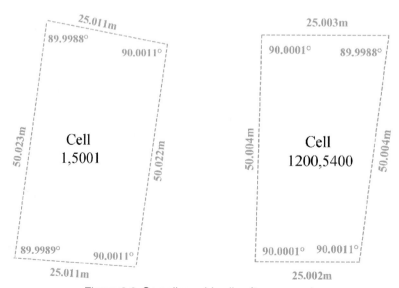

Figure 6.8 Sampling grid cells after conversion.

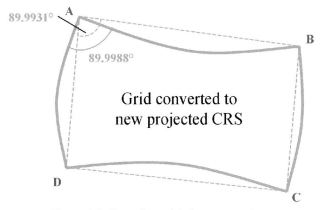

Figure 6.9 **Sampling grid after conversion.**

Retention of grid on another projection

Can we retain the geometry of the original grid in a different projected CRS to that in which the grid was defined? The simple answer is, in general, 'no'. There are two options. The first is to apply the grid definition to our coordinate reference system. This will result in the retention of the shape and size of the grid, but there is an infinite number of ways that we can position the grid. Figure 6.10 shows two possibilities – by retaining cell A at the same point as in the original definition and with the side AB orientated towards point B in the target CRS, in which case cells away from A will be increasingly out of position, or by finding a best fit position for the corners ABCD. In these and all other possibilities, it is impossible to have all cells fall in their correct position in the target CRS.

The second possibility, and the only one that will honour the original data, is to resample the data on the new projected CRS.

Any other attempt to fit the grid to the new projection will be applying an affine or polynomial transformation (section 4.5) and not honour the geometry of the original grid.

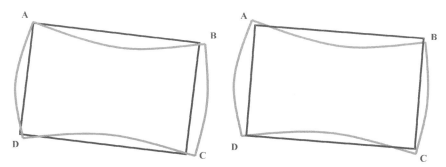

Figure 6.10 **The grid laid onto the new projection. The blue rectangle represents the same size and shape as the original definition, the green polygon is that shape converted to the new projection.**

6.5   Creating overlays in Google Earth™

This case study will examine the problems that ensue when overlaying two data sets that are in different projected coordinate reference systems. It will use the specific example of Google Earth™, but users of other systems that perform similar functions will be able to draw appropriate comparisons.

We have a raster data set showing a geological map of California (Figure 6.11). We wish to create an overlay in Google Earth™ in order to explore visually the correlation between rock types and terrain.

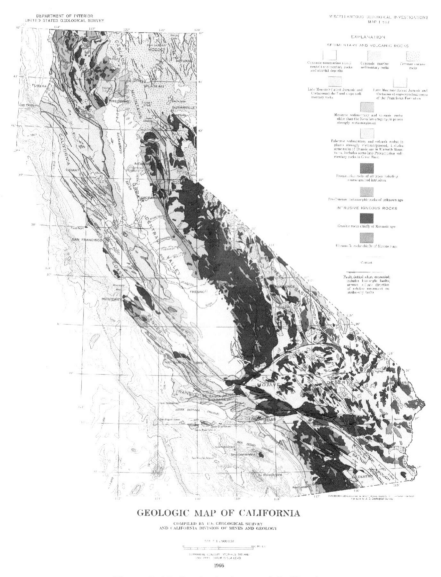

Figure 6.11 Geological map of California.

It will be recalled from section 4.5.6 that Google Earth™ expects any overlay image to be imported in the cylindrical equidistant projection. If our geological map were a properly geo-referenced data set, then it would be a simple step to re-project it in this way. Unfortunately, we only have the map as an un-referenced image file, and by inspection it is clearly not in a cylindrical equidistant projection. After importing the image, and applying the affine manipulations that the system permits us, the best that we can do is as shown in Figure 6.12.

Trying to make it fit better in one place makes it worse in others, and whatever happens we can see by inspection of the coastline and border that we end up with mismatches of around 10–30 km. This is as predicted by equation 4.48.

Using just Google Earth™, we cannot really improve on this, and it is necessary to turn to an alternative software package that permits the manipulation of raster data. One possible approach would be to use a 'rubber sheeting' or polynomial type approach, in which graticule points on the original map are dragged to the positions that they should be in on the cylindrical equidistant projection. The disadvantage of this is that it is a rather unstable procedure, and achieving a match at selected

Figure 6.12 Best fit of the overlay from an affine transformation.
(Courtesy of: TerraMetrics, Inc. www. truearth.com and
Europa Technologies Ltd. www.europa-tech.com)

control points would not necessarily mean that the two data sets fitted in between. A considerable amount of work would have to be done to establish tie points at a sufficiently dense coverage.

A better approach is to use a system that can be made to 'understand' the existing projection, before re-projecting it in the required cylindrical equidistant. We used the ESRI system ArcGIS for this, but ran up against the problem that no information on the existing projection is printed on the map.

So what is the projection? By inspection of Figure 6.11, the meridians are straight convergent lines and the parallels are curved. Since the angle between any two meridians is less than their difference in longitude, this is clearly a conic projection. The meridian at 120°W runs vertically up and down the image and so this can be taken as the central meridian or longitude of origin – this may not strictly be true, as an alternative would have been to choose another meridian and then rotate the map, but for our purposes this is not important as the geometry of the projection ends up being the same. We measured the angle between this meridian and 123°W by measuring the offset at the top and bottom of the image and computing it from trigonometry as 1.888° (this is more accurate than a protractor would be able to measure). Then from

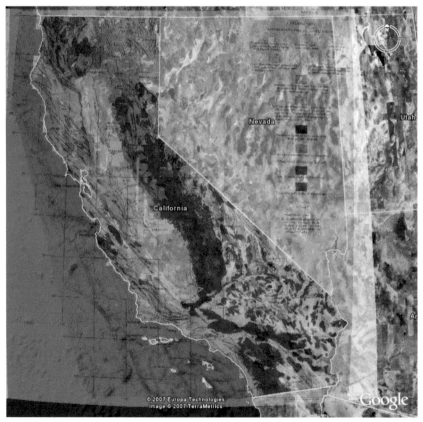

Figure 6.13 **Overlay re-projected to cylindrical equidistant projection and then imported into Google Earth™. (Courtesy of: TerraMetrics, Inc. www. truearth.com and Europa Technologies Ltd. www.europa-tech.com)**

the general formula for the convergence on conic map projections (equation 3.31), this ratio of 1.888°/3° implies a standard parallel of 39°N (the number comes out quite close to an integer value, which gives us some confidence that we have found the right one).

The exact type of projection is more problematic, as the scale is too small to measure easily how the distance between the parallels changes with latitude. However, if it is not easy to tell then it probably does not matter that much, and so as the most likely choice we opt for Lambert Conformal Conic (LCC). We can either use the single standard parallel of 39°N, or we can opt for two standard parallels that give this average, such as 34°N and 44°N. These are the most important features of the projection – others, such as false eastings and northings – would not affect the shape of the projection in any way, and so we assign arbitrary values.

Have we found the exact projection used by the US Geological Survey to create the map? Probably not – but we have devised a projection with very similar geometry, and are therefore reasonably confident of our ability to geo-reference it.

We then set up our chosen LCC projection in ArcGIS™ using the parameters derived and plot a few key graticule points around the edge of the area. The image of the original map is then imported to this layer and shifted and scaled to fit onto the tie points – this works because the imported image and the tie points are now in the same projection (or close approximations) and therefore have the same essential geometry.

We now have a geo-referenced image, and this can then be re-projected into the cylindrical equidistant projection. We now import this new image into Google Earth™, and the result is shown as Figure 6.13. The scale of the original map from which this image was digitised was 1:2.5M, and we now have a match that is consistent with the accuracy that this implies.

APPENDIX A
TERMINOLOGY

Entries in this table are listed by ISO term in alphabetical order. Colloquial terms are given against the ISO equivalent and are also inserted alphabetically within the table.

ISO 19111 term	Colloquial term	Comment
affine coordinate system		Coordinate system in Euclidean space with straight axes that are not necessarily mutually perpendicular.
	altitude	Height where the chosen reference surface is mean sea level. See *height*.
Cartesian coordinate system		Coordinate system with 2 or 3 mutually perpendicular straight axes.
	chart datum	Vertical reference surface used as the basis for reporting water depth.
compound coordinate reference system (CCRS)		Coordinate reference system using at least two independent coordinate reference systems. Coordinate reference systems are independent of each other if coordinate values in one cannot be converted or transformed into coordinate values in the other.
	conversion	See *coordinate conversion.*
coordinate	ordinate, co-ordinate	One of a sequence of n numbers designating the position of a point in n-dimensional space.
coordinate conversion	conversion	Change of coordinates from one CRS to another in which the two CRSs are based on the same datum. Subset of coordinate operation.
coordinate operation		Change of coordinates, based on a one-to-one relationship, from one coordinate reference system to another. Superset of coordinate conversion and coordinate transformation.
coordinate operation parameter	parameter	Name of a variable used in a conversion or transformation.
coordinate operation parameter value	parameter value	Value assigned to a coordinate operation parameter.

ISO 19111 term	Colloquial term	Comment
coordinate reference system (CRS)	coordinate system	Coordinate system that has its position, scale, and orientation defined with respect to an object; this is accomplished through a datum. In the context of this book, the object is usually the Earth. However, it could be an object moving relative to the Earth, for example an aeroplane carrying an aerial camera.
		Care with colloquial usage! Coordinate System is defined in ISO 19111, but with a narrower definition than that for Coordinate Reference System.
coordinate set		Collection of coordinate tuples related to the same coordinate reference system (see below).
coordinate system (CS)		Set of rules for defining how coordinates are assigned to points. Coordinates describing locations on the Earth or some other object are referenced to a Coordinate Reference System (see above).
	coordinate system	See *coordinate reference system*.
coordinate transformation	transformation	Change of coordinates from one CRS to another in which the CRSs are based on different datums. Subset of coordinate operation.
coordinate tuple		Ordered list of coordinates describing the position of one location in n-dimensional space where $n > 1$. For example, the values of latitude and longitude for one point.
datum		A datum defines the position of the origin, the scale and the orientation of the axis or axes of a coordinate system with respect to an object.
	datum transformation	A misnomer. It is coordinates, not datums, that the transformation acts upon. See coordinate transformation.
depth		Distance of a point from a chosen reference surface measured downward along a line perpendicular to that surface. A depth above the reference surface will have a negative value.
easting		Distance in a coordinate system, eastwards (positive) or westwards (negative) from a north-south reference.
	elevation (US)	See *height*.
ellipsoid	spheroid	An ellipsoid is a closed surface formed by the rotation of an ellipse about a main axis whose plane sections are either ellipses or circles. A spheroid is a closed surface that differs only slightly from that of a sphere; its plane sections are not necessarily ellipses or circles – they can be irregular.
		The figure of the Earth is modelled as an ellipsoid that differs only slightly from a sphere – hence the colloquial usage of spheroid and ellipsoid as synonyms.

(Continued)

ISO 19111 term	Colloquial term	Comment
ellipsoidal coordinate system	geodetic coordinate system	Coordinate system in which position is specified by geodetic latitude, geodetic longitude and (in the three-dimensional case) ellipsoidal height.
ellipsoidal coordinates	geodetic coordinates, geographic coordinates	Astronomic or geodetic latitude and longitude (and in the geodetic 3D case, ellipsoidal height).
ellipsoidal height (h)	geodetic height	Distance of a point from the ellipsoid measured along the perpendicular from the ellipsoid to this point; positive if upwards or outside of the ellipsoid. Only used as part of a three-dimensional ellipsoidal coordinate system and never on its own.
engineering coordinate reference system	local coordinate (reference) system	Coordinate reference system based on an engineering datum. For example, a grid for a construction site or an industrial plant, or a coordinate reference system local to a ship or an orbiting spacecraft.
engineering datum	local datum	Datum describing the relationship of a coordinate system to a local reference.
	geocentric coordinate reference system	Geodetic coordinate reference system with origin at the centre of the Earth. Generally taken to be a geocentric Cartesian coordinate system, but in some space applications may use a spherical coordinate system.
	geocentric coordinates, geocentric Cartesian coordinates	3D Cartesian coordinates referenced to the centre of the Earth. Note: ISO 19111 has no breakdown for types of coordinates. It distinguishes between types of coordinate system by the geometric properties of the coordinate space spanned and the CS's axes. It also distinguishes between types of CRS (by the type of datum associated with the CRS).
geodetic coordinate reference system		Coordinate reference system based on a geodetic datum.
	geodetic coordinates, geographic coordinates	Geodetic latitude, geodetic longitude and (in the three-dimensional case) ellipsoidal height. Equivalent to ISO 19111 ellipsoidal coordinate system. Note: ISO 19111 has no breakdown for types of coordinates. It distinguishes between types of coordinate system by the geometric properties of the coordinate space spanned and the CS's axes. It also distinguishes between types of CRS (by the type of datum associated with the CRS).
geodetic datum		Datum describing the relationship of a 2- or 3-dimensional coordinate system to the Earth.
geodetic latitude (φ)	latitude, ellipsoidal latitude	Angle from the equatorial plane to the perpendicular to the ellipsoid through a given point, northwards treated as positive.

ISO 19111 term	Colloquial term	Comment
geodetic longitude (λ)	longitude, ellipsoidal longitude	Angle from the prime meridian plane to the meridian plane of a given point, eastwards treated as positive.
	geographic coordinate reference system	Coordinate reference system using latitude and longitude. Could be astronomic, but usually geodetic in which case may optionally include ellipsoidal height as a third dimension.
geoid		Equipotential surface of the Earth's gravity field, which is everywhere perpendicular to the direction of gravity and which best fits mean sea level either locally or globally.
geoid-ellipsoid separation (N)	geoid height, geoid undulation	Height of the geoid above the ellipsoid. See section 2.3.4.2.
gravity-related height (H)	height	Height dependent on the Earth's gravity field, in particular, orthometric height or normal height, which are both approximations of the distance of a point above mean sea level.
height		Distance of a point from a chosen reference surface measured upward along a line perpendicular to that surface. A height below the reference surface will have a negative value. Superset of ellipsoidal height and gravity-related height.
		When used colloquially, generally taken to mean gravity-related height. Might occasionally mean ellipsoidal height.
image coordinate reference system		Coordinate reference system based on an image datum.
image datum		Engineering datum which defines the relationship of a coordinate system to an image
	latitude	See *geodetic latitude*.
	local datum	See *engineering datum* and *engineering CRS*.
	longitude	See *geodetic longitude*.
map projection	projection	Coordinate conversion from an ellipsoidal coordinate system to a plane.
mean sea level (MSL)		Average level of the surface of the sea over all stages of tide and seasonal variations. Mean sea level in a local context normally means mean sea level for the region calculated from observations at one or more points over a given period of time. Mean sea level in a global context differs from a global geoid by not more than 2 m.

(Continued)

ISO 19111 term	Colloquial term	Comment
meridian		Intersection of an ellipsoid by a plane containing the shortest axis of the ellipsoid.
northing		Distance in a coordinate system, northwards (positive) or southwards (negative) from an east-west reference line.
	ordinate	See *coordinate*.
prime meridian	zero meridian	Meridian from which the longitudes of other meridians are quantified.
projected coordinate reference system		Coordinate reference system derived from a two-dimensional geodetic coordinate reference system by applying a map projection.
	projected coordinates	2D Cartesian coordinates referenced to a projected CRS. Note: ISO 19111 has no breakdown for types of coordinates. It distinguishes between types of coordinate system by the geometric properties of the coordinate space spanned and the CS's axes. It also distinguishes between types of CRS (by the type of datum associated with the CRS).
	projection	See *map projection*.
	spheroid	See *ellipsoid*.
	transformation	See *coordinate transformation*.
vertical coordinate reference system		One-dimensional coordinate reference system based on a vertical datum.
vertical coordinate system		One-dimensional coordinate system used for gravity-related height or depth measurements.
vertical datum		Datum describing the relation of gravity-related heights or depths to the Earth. In most cases the vertical datum will be related to mean sea level. Ellipsoidal heights are treated as related to a three-dimensional ellipsoidal coordinate system referenced to a geodetic datum. Vertical datums include sounding datums (used for hydrographic purposes), in which case the heights may be negative heights or depths.
	zero meridian	See *prime meridian*.

APPENDIX B
COMPUTATIONS WITH
SPHERICAL COORDINATES

This appendix is a summary of the formulae that are used to determine distances and azimuths from spherical coordinates.

The shortest route between two points on the sphere is referred to as the *great circle*. This can be defined by the intersection with the sphere of a plane that passes through both the points and the centre of the sphere.

Let the coordinates of two points, A and B, be given as (φ_A, λ_A) and (φ_B, λ_B), respectively.

The great circle distance L_{AB} between the two points is given as:

$$\cos L_{AB} = \sin\varphi_A \sin\varphi_B + \cos\varphi_A \cos\varphi_B \cos\Delta\lambda \qquad (B.1)$$

where $\Delta\lambda$ is the difference in longitude between the two points:

$$\Delta\lambda = \lambda_B - \lambda_A \qquad (B.2)$$

This gives an answer in angular units, which may be converted to distance by expressing the angle in radians and multiplying by an appropriate value for the radius of the spherical Earth:

$$L_{(km)} = 6371 \frac{\pi}{180} L_{(degrees)} \qquad (B.3)$$

The azimuth of the point B from A is defined as the clockwise angle between the meridian at A and the great circle, and is given by the expression:

$$\cot A_{AB} = \frac{\cos\varphi_A \tan\varphi_B - \sin\varphi_A \cos\Delta\lambda}{\sin\Delta\lambda} \qquad (B.4)$$

APPENDIX C
BASIC GEOMETRY OF
THE ELLIPSOID

C.1 Introduction

The ellipsoid is the basic reference surface for the definition of geographic coordinates. With the use of GNSS for most surveys that cover a large area, the need to carry out computations using geodetic coordinates has greatly diminished, and most local surveys will be carried out using projected coordinates that have been suitably corrected.

One of the few remaining uses for computations on the ellipsoid is in relation not to actual observations, but to the establishment of boundaries: this may be, for example, between states or between oil concessions. In these situations it may be necessary, for example, to compute the coordinates of a boundary line between two defining points that lie several hundred kilometres apart. In such situations, the use of a projection is inappropriate, and neither can the computations be carried out in geocentric Cartesian coordinates.

What follows is by no means a comprehensive discussion of all aspects of geometrical geodesy, but is a summary of some of the most useful concepts and formulae. A fuller treatment may be found in references such as Bomford (1980).

C.2 Radii of curvature of the ellipsoid

The sphere has a single radius, usually designated by the symbol R. For the ellipsoid, the *radius of curvature* is not the same as the distance from the centre of the figure. In fact it is a function both of the latitude and the direction that is being considered. Its values in the cardinal directions are referred to as v, the radius of curvature in the prime vertical (or east-west direction) and ρ, the radius of curvature in the meridional section (or north-south direction). These values are given by the expressions:

$$v = \frac{a}{(1 - e^2 \sin^2\varphi)^{1/2}} \tag{C.1}$$

$$\rho = \frac{a(1 - e^2)}{(1 - e^2 \sin^2\varphi)^{3/2}} \tag{C.2}$$

If required, the radius of curvature R_α in a general direction α (defined as a clockwise azimuth from true north) can be computed from:

$$\frac{1}{R_\alpha} = \frac{\cos^2\alpha}{\rho} + \frac{\sin^2\alpha}{v} \tag{C.3}$$

C.3 Normal sections and geodesics

The azimuth from point A to point B (both having been projected onto the surface of the ellipsoid, if necessary) is defined firstly with reference to the plane that contains

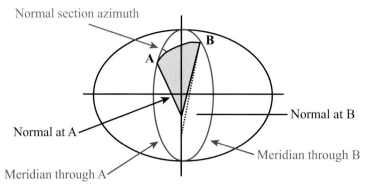

Normal section azimuth

A

B

Normal at B

Normal at A

Meridian through B

Meridian through A

Figure C.1 **Normal section at A to the point B.**

the normal to the ellipsoid at point A, and both points A and B. The angle between this plane and the meridional section at A, when measured clockwise from north, defines the normal section azimuth from A to B, as shown in Figure C.1.

The normal section from A to B is not, in general, the same plane as that from B to A. The intersection of these two planes with the surface of the ellipsoid therefore establishes two separate lines. Except in special cases, neither of these two lines is actually the shortest route between the two points.

The shortest route between two points on the surface of the ellipsoid is called the geodesic. It is a complex line that cannot be described by the intersection of a plane with the surface of the ellipsoid. As shown in Figure C.2, it will lie in between the two normal sections with an initial azimuth at A that is closer to the normal section from A than the normal section from B in the ratio 2:1.

This property also applies in reverse, in that the azimuth of the geodesic at the point B will be closer to the normal section from B than to the normal section from A in the same ratio. The maximum distance between the two normal sections is a function of the distance between the two points, and ranges from a few centimetres for lines up to 100 km, to several metres for lines up to 1000 km.

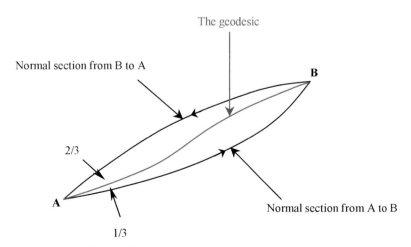

The geodesic

Normal section from B to A

B

2/3

A

Normal section from A to B

1/3

Figure C.2 **Normal sections and the geodesic.**

C.4 Forward computation of coordinates

Several formulae may be used to determine the coordinates of a point given the coordinates of an initial point and the distance and azimuth. Some of them are less complex than the one quoted here, but have limited accuracy over longer distances.

Given the coordinates of point A (φ_A, λ_A), and a normal section azimuth (A) and distance (L), it is required to find the coordinates of the point B, such that:

$$\varphi_B = \varphi_A + \Delta\varphi \tag{C.4}$$

$$\lambda_B = \lambda_A + \Delta\lambda \tag{C.5}$$

Clarke's 'best formulae' may be used for distances up to 1200 km with an accuracy of within 1/25 of a part per million (Bomford 1980). These are:

$$\varepsilon = \frac{e^2}{1 - e^2} \tag{C.6}$$

$$r_2' = -\varepsilon \, \cos^2\varphi_A \, \cos^2 A \tag{C.7}$$

$$r_3' = 3\varepsilon(1 - r_2') \; \cos\varphi_A \, \sin\varphi_A \, \cos A \tag{C.8}$$

$$\theta = \frac{L}{v_A} - \frac{r_2'\,(1 + r_2')}{6}\left(\frac{L}{v_A}\right)^3 - \frac{r_3'\,(1 + 3r_2')}{24}\left(\frac{L}{v_A}\right)^4 \tag{C.9}$$

$$\frac{v_A}{r} = 1 - \frac{r_2'}{2}\,\theta^2 - \frac{r_3'}{2}\,\theta^3 \tag{C.10}$$

$$\sin\psi = \sin\varphi_A \, \cos\theta + \cos\varphi_A \, \cos A \, \sin\theta \tag{C.11}$$

Then:

$$\sin\Delta\lambda = \sin A \, \sin\theta \, \sec\psi \tag{C.12}$$

$$\tan\varphi_B = (1 + \varepsilon)\left\{1 - e^2\left(\frac{v_A}{r}\right)\frac{\sin\varphi_A}{\sin\psi}\right\}\tan\psi \tag{C.13}$$

All terms not specifically defined here are given in Chapter 2.

C.5 Reverse computation of azimuth

A normal section azimuth, A_{AB}, between two points A (φ_A, λ_A) and B (φ_B, λ_B), which are of known coordinates, may be computed as follows:

$$\tan A_{AB} = \frac{-\Delta X \sin\lambda_A + \Delta Y \cos\lambda_A}{-\Delta X \sin\varphi_A \cos\lambda_A - \Delta Y \sin\varphi_A \sin\lambda_A + \Delta Z \cos\varphi_A} \tag{C.14}$$

The geocentric Cartesian coordinates of A and B are found from the equations in section 2.3.4, and the convention for the differences is:

$$\Delta X = X_B - X_A \tag{C.15}$$

with similar terms for Y and Z.

C.6   Determination of points on the geodesic

It is not unusual that a boundary between two areas (mineral concessions or neighbouring states) is defined as the geodesic between two points of given coordinates (in a certain coordinate reference system). It will then be necessary to determine the coordinates of points at given intervals along the geodesic.

The fact that the geodesic is not a plane curve means that the methods for computing it are very complicated, usually involving the numerical expansion of an integral. Sharma (1966) gives a rigorous method for this. As an alternative, it is possible within a limited accuracy to compute points on the geodesic by interpolating between the normal sections. A suggested method follows.

1. Given the coordinates of the two end points, A and B, a distance L apart, compute the normal section azimuths from each end by the use of equation C.14.

2. Compute the coordinates of a point on the normal section from A to B at a distance D using equations C.4 to C.13. Let this point be called P_A.

3. Compute the coordinates of a point on the normal section from B to A at a distance (L-D), using the same formulae. Let this point be called P_B. The situation is illustrated in Figure C.3.

It is now possible to interpolate the coordinates of the point on the geodesic from the coordinates of the two points found so far if we bear in mind that the geodesic is nearer to the normal section AB than the section BA in the ratio 2:1 at the point A, half way in between the two at the mid-point, and nearer BA than AB in the ratio 2:1 at the point B.

Making an assumption of a linear change in the ratio, this gives the coordinates of the geodesic point (φ_G, λ_G) as:

$$\lambda_G = \lambda_{P_A} + \left(\lambda_{P_B} - \lambda_{P_A}\right)\left[\frac{1+\dfrac{D}{L}}{3}\right] \tag{C.16}$$

$$\varphi_G = \varphi_{P_A} + \left(\varphi_{P_B} - \varphi_{P_A}\right)\left[\frac{1+\dfrac{D}{L}}{3}\right] \tag{C.17}$$

This technique has an estimated accuracy of within 10 cm at distances up to 1000 km.

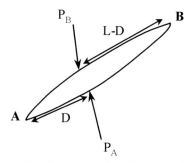

Figure C.3 Approximation of the geodesic.

APPENDIX D
THE MOLODENSKY EQUATIONS

The direct transformation from one geographic coordinate reference system to another is covered in section 4.4.2. The equations are due to Molodensky (Stansell 1978) and in their full form are given as:

$$\Delta\varphi" =$$

$$\frac{-\Delta X \sin\varphi \cos\lambda - \Delta Y \sin\varphi \sin\lambda + \Delta Z \cos\varphi + \Delta a \dfrac{(ve^2 \sin\varphi \cos\varphi)}{a} + \Delta f (\rho \dfrac{a}{b} + v \dfrac{b}{a}) \sin\varphi \cos\varphi}{(\rho + h) \sin 1"} \tag{D.1}$$

$$\Delta\lambda" = \frac{-\Delta X \sin\lambda + \Delta Y \cos\lambda}{(v + h) \cos\varphi \sin 1"} \tag{D.2}$$

$$\Delta h = \Delta X \cos\varphi \cos\lambda + \Delta Y \cos\varphi \sin\lambda + \Delta Z \sin\varphi - \Delta a \frac{a}{v} + \Delta f \frac{b}{a} v \sin^2 \varphi \tag{D.3}$$

The changes $\Delta\varphi, \Delta\lambda, \Delta h$ that result from these are to be added to the source coordinates to arrive at the target coordinates. The angular units are seconds of arc and the height is expressed in metres (as are all required linear terms such as ellipsoid parameters a and b). The parameter changes $\Delta a, \Delta f$ are the differences that result from subtracting the parameters of the *source* system from those of the *target* system.

The terms $\Delta X, \Delta Y, \Delta Z$ have the same meaning as when they were introduced in section 4.3 covering geocentric transformations. The parameters v and ρ are as defined in equations C.1 and C.2.

In equations D.1 to D.3, all required parameter terms such as a, b, e, ρ, and v are computed for the *source* coordinate reference system, as are the input values of latitude, longitude, and ellipsoidal height.

The Abridged Molodensky formulae simplify the expressions for computational efficiency, and are given by equations D.4 to D.6.

$$\Delta\varphi" = \frac{-\Delta X \sin\varphi \cos\lambda - \Delta Y \sin\varphi \sin\lambda + \Delta Z \cos\varphi + (a \Delta f + f \Delta a) \sin 2\varphi}{\rho \sin 1"} \tag{D.4}$$

$$\Delta\lambda" = \frac{-\Delta X \sin\lambda + \Delta Y \cos\lambda}{v \cos\varphi \sin 1"} \tag{D.5}$$

$$\Delta h = \Delta X \cos\varphi \cos\lambda + \Delta Y \cos\varphi \sin\lambda + \Delta Z \sin\varphi + (a \Delta f + f \Delta a) \sin^2\varphi - \Delta a \tag{D.6}$$

The units and conventions used for these expressions are the same as for the full Molodensky formulae.

APPENDIX E
DETERMINATION OF
TRANSFORMATION PARAMETER
VALUES BY LEAST SQUARES

E.1 Introduction and least squares terminology

The purpose of this appendix is to bring together the mathematical processes for determining the parameters of any transformation model by least squares, given the coordinates in two different systems of a sufficient number of points.

A basic familiarity with the least squares process has to be assumed, but a recapitulation of the essential equations will be given here to establish the terminology. The notation in this appendix follows that of Allan (1997b; 2007), and the specific application to transformation problems is based on the approach of Cooper (1987).

The aim of the least squares process is, in general, to determine best estimates of a set of *unobserved parameters*, $\mathbf{x}$, from a set of *observed parameters*, $\mathbf{s}$. The bold type indicates that we are dealing here with *vectors*, containing several parameters. For this particular application, the unobserved parameters will usually be the parameters of the transformation, and the observed parameters will be coordinate values.

The relationship between $\mathbf{x}$ and $\mathbf{s}$ is expressed via a functional relationship, F, where:

$$F(\hat{\mathbf{x}}; \hat{\mathbf{s}}) = 0 \qquad\qquad (E.1)$$

and where $\hat{\mathbf{x}}$ represents the best estimate of $\mathbf{x}$ and $\hat{\mathbf{s}}$ represents the best estimate of $\mathbf{s}$. Since in general we are dealing with non-linear transformations, the functional relationship is linearised as follows:

$$F(\hat{\mathbf{x}};\hat{\mathbf{s}}) = F(\dot{\mathbf{x}};\dot{\mathbf{s}}) + \frac{\partial F}{\partial \mathbf{x}}\,\mathbf{dx} + \frac{\partial F}{\partial \mathbf{s}}\,\mathbf{ds} \qquad\qquad (E.2)$$

where $\dot{\mathbf{x}}$ and $\dot{\mathbf{s}}$ represent provisional values of the unobserved and observed parameters, respectively, and are related to the best estimates through the expressions:

$$\hat{\mathbf{x}} = \dot{\mathbf{x}} + \mathbf{dx} \qquad\qquad (E.3)$$

and

$$\hat{\mathbf{s}} = \dot{\mathbf{s}} + \mathbf{ds} \qquad\qquad (E.4)$$

In addition, the best estimates of the observed parameters are related to the observations themselves, $\overset{\circ}{\mathbf{s}}$, by the expression:

$$\hat{\mathbf{s}} = \overset{\circ}{\mathbf{s}} + \mathbf{v} \qquad\qquad (E.5)$$

where the quantities **v** are described as the residuals. Equations E.3 to E.5 can then be combined to express the vector **ds** in terms independent of the best estimates as:

$$\mathbf{ds} = \overset{\circ}{\mathbf{s}} - \dot{\mathbf{s}} + \mathbf{v} \qquad (E.6)$$

$$= \mathbf{l} + \mathbf{v} \qquad (E.7)$$

In equation E.2, the functional relationship, F, will evaluate as zero when it is applied to the best estimates of the observed and the unobserved parameters, as this is the way that the relationship was first introduced in E.1. It is also the case that the relationship will be zero when applied to the provisional values of the observed and unobserved parameters, provided that they are given as a consistent set. Thus, equation E.2 can be simplified to:

$$\frac{\partial F}{\partial \mathbf{x}} \, \mathbf{dx} + \frac{\partial F}{\partial \mathbf{s}} \, \mathbf{ds} = 0 \qquad (E.8)$$

For convenience, **dx** will henceforward be referred to simply as **x**, the vector of unknowns that must be added to the provisional values to obtain the best estimates. Similarly, the vector **ds** is referred to as **s**, and is split into its component parts **l** + **v**. Thus the equation now becomes:

$$\mathbf{Ax} + \mathbf{Cv} = -\mathbf{Cl} = \mathbf{b} \qquad (E.9)$$

where

$$\mathbf{A} = \frac{\partial F}{\partial \mathbf{x}} \qquad (E.10)$$

and

$$\mathbf{C} = \frac{\partial F}{\partial \mathbf{s}} \qquad (E.11)$$

A and **C** are termed *Jacobian matrices*, and represent the partial derivatives of the functional relationships with respect to the unobserved and observed parameters, respectively. More concisely, they are often called the *design matrices*. Their form will be discussed in detail in relation to the different transformation models in the following sections, but it should be pointed out here that they have been derived from a linearisation that has the provisional values of the observed and unobserved parameters as its starting point. In evaluating the matrices, the provisional values of these parameters should therefore be used.

In some situations it is possible to express a problem in such a way that the design matrix **C** reduces to the identity matrix, **I**. In such cases, equation E.8 can be re-arranged and expressed in the form:

$$\mathbf{Ax} = \mathbf{l} + \mathbf{v} \qquad (E.12)$$

If the observation equations are given in this form, then it can be shown (Allan 1997b; 2007) that the solution according to the weighted sum of the squares of the residuals, **v**, is:

$$\mathbf{x} = (\mathbf{A}^{\mathsf{T}} \, \mathbf{WA})^{-1} \, (\mathbf{A}^{\mathsf{T}} \mathbf{Wl}) \qquad (E.13)$$

where **W** is an appropriate weight matrix.

Alternatively, if the more general form of equation E.9 is used, then it can be shown (Allan 1997b; 2007) that the solution now has the form:

$$\mathbf{x} = (\mathbf{A}^T (\mathbf{CW}^{-1}\mathbf{C}^T)^{-1} \mathbf{A})^{-1} (\mathbf{A}^T (\mathbf{CW}^{-1}\mathbf{C}^T)^{-1} \mathbf{b}) \qquad (E.14)$$

In the sections that follow, several different forms of transformation models will be considered. For each of these, the functional relationship between observed and unobserved parameters will first be defined, as will the weight matrix. The form of the design matrices will then be derived. After this, the solution proceeds by the application of either equation E.13 or equation E.14.

Before proceeding to a consideration of the application to different transformation models, however, further consideration should be given to the formation of the vector $\mathbf{b}$ in equation E.9. The implication of the way that it was derived is that it is first necessary to form the vector $\mathbf{l}$ from knowledge of the provisional values of the observed parameters. Whilst for many problems in least squares this is not a problem, in the area of coordinate transformations this is often an unwieldy procedure. An alternative derivation of equation E.9 will therefore be given (as in Cross, 1983) that obviates the need to form provisional values of the observations.

This starts with the alternative definition:

$$F(\hat{\mathbf{x}};\hat{\mathbf{s}}) = F(\dot{\mathbf{x}};\overset{\circ}{\mathbf{s}}) + \frac{\partial F}{\partial \mathbf{x}}\,d\mathbf{x} + \frac{\partial F}{\partial \mathbf{s}}\mathbf{v} \qquad (E.15)$$

which follows from the fact that the residuals, $\mathbf{v}$, are defined as the differences between the observations and the best estimates. This can then be simplified to:

$$\mathbf{A}\mathbf{x} + \mathbf{C}\mathbf{v} = -F(\dot{\mathbf{x}};\overset{\circ}{\mathbf{s}}) = \mathbf{b} \qquad (E.16)$$

Therefore, $\mathbf{b}$ can be found by substituting the observations (observed coordinates) and the provisional values of the unobserved parameters (transformation parameters) into the functional relationship. It should also be noted that, with the linearisation now starting from the provisional values of the unobserved parameters and the observed values of the observed parameters, it is these values that should be used in evaluating the matrices $\mathbf{A}$ and $\mathbf{C}$. This is not always stated explicitly in the derivations that follow, as it would make the notation rather unwieldy, but it is implicit throughout.

E.2 Two dimensional transformations of Cartesian coordinates

E.2.1 The Similarity transformation

A set of coordinates in the source system (X_S, Y_S) is to be transformed into the target system (X_T, Y_T). The two-dimensional four-parameter Similarity transformation method was introduced in section 4.5.3 as:

$$X_T = a_0 + aX_S + bY_S \qquad (E.17)$$

$$Y_T = b_0 - bX_S + aY_S \qquad (E.18)$$

The first step is to re-cast these expressions into the form of a set of functional relationships in the form required by equation (E.1).

$$F_1 = a_0 + aX_S + bY_S - X_T \qquad (E.19)$$

$$F_2 = b_0 + bX_S + aY_S - Y_T \qquad (E.20)$$

It will be recalled from section 4.5.3 that the parameters used here can be related to the geometrical parameters of scale factor (μ) and rotation angle (α) by:

$$a = \mu \cos\alpha \qquad (E.21)$$

$$b = \mu \sin\alpha \qquad (E.22)$$

The terms a_0 and b_0 are equivalent to translations in the X and Y axes respectively. Therefore in solving for the parameters a and b the geometrical parameters can subsequently be recovered from:

$$\mu = \sqrt{a^2 + b^2} \qquad (E.23)$$

$$\alpha = \tan^{-1}\left(\frac{b}{a}\right) \qquad (E.24)$$

The situation is further simplified if it is possible to say that the coordinates (X_S, Y_S) are effectively constants, and may be regarded as being without error. In this case, the only observed parameters are the coordinates (X_T, Y_T), while the unobserved parameters are a, b, ΔX, and ΔY.

The design matrices **A** and **C** then have the form:

$$\mathbf{A} = \begin{vmatrix} \dfrac{\partial F_1}{\partial a} & \dfrac{\partial F_1}{\partial b} & \dfrac{\partial F_1}{\partial a_0} & \dfrac{\partial F_1}{\partial b_0} \\[2mm] \dfrac{\partial F_2}{\partial a} & \dfrac{\partial F_2}{\partial b} & \dfrac{\partial F_2}{\partial a_0} & \dfrac{\partial F_2}{\partial b_0} \end{vmatrix} \qquad (E.25)$$

$$\mathbf{C} = \begin{vmatrix} \dfrac{\partial F_1}{\partial X_T} & \dfrac{\partial F_1}{\partial Y_T} \\[2mm] \dfrac{\partial F_2}{\partial X_T} & \dfrac{\partial F_2}{\partial Y_T} \end{vmatrix} \qquad (E.26)$$

These are then evaluated as:

$$\mathbf{A} = \begin{pmatrix} X_S & Y_S & 1 & 0 \\ Y_S & -X_S & 0 & 1 \end{pmatrix} \qquad (E.27)$$

$$\mathbf{C} = \begin{pmatrix} -1 & 0 \\ 0 & -1 \end{pmatrix} \qquad (E.28)$$

As the matrix **C** found above has the form –**I**, it can be seen that the general form of the observation equations given in equation E.9 can be reduced to the special case of equation E.12.

In fact, the matrices determined in equations E.27 and E.28 are only a part of the full picture, as these have been derived for only one point. In general, there will be a similar matrix for each point that is common to the two coordinate systems. Denoting the partial matrix by the symbol $\mathbf{A}_i$, where:

$$\mathbf{A}_i = \begin{pmatrix} X_{Si} & Y_{Si} & 1 & 0 \\ Y_{Si} & -X_{Si} & 0 & 1 \end{pmatrix} \tag{E.29}$$

and (X_{Si}, Y_{Si}) are the coordinates of the ith point, the full design matrix for n points is made up as follows:

$$\mathbf{A} = \begin{pmatrix} \mathbf{A}_1 \\ \mathbf{A}_2 \\ \vdots \\ \mathbf{A}_n \end{pmatrix} \tag{E.30}$$

which is a (2n × 4) matrix.

The more general form of the transformation is the situation where it is not possible to assume that one set of coordinates is correct, and instead both (X_T, Y_T) and (X_S, Y_S) are assumed to be observed data with associated weights.

The functional relationships are the same as before, but they will now be written in a form that makes it explicit that they are two equations out of a total set of 2n. Thus:

$$F_{1i} = aX_{Si} + bY_{Si} + a_0 - X_{Ti} \tag{E.31}$$

$$F_{2i} = -bX_{Si} + aY_{Si} + b_0 - Y_{Ti} \tag{E.32}$$

As before, the sub matrix $\mathbf{A}_i$ is determined by differentiating with respect to the unobserved parameters, and is exactly the same as previously:

$$\mathbf{A}_i = \begin{pmatrix} X_{Si} & Y_{Si} & 1 & 0 \\ Y_{Si} & -X_{Si} & 0 & 1 \end{pmatrix} \tag{E.33}$$

The sub-matrix $\mathbf{C}_i$ is now different to before, however, as it includes the derivatives with respect to all observed parameters, thus:

$$\mathbf{C}_i = \begin{vmatrix} \dfrac{\partial F_{1i}}{\partial X_{Si}} & \dfrac{\partial F_{1i}}{\partial Y_{Si}} & \dfrac{\partial F_{1i}}{\partial X_{Ti}} & \dfrac{\partial F_{1i}}{\partial Y_{Ti}} \\ \dfrac{\partial F_{2i}}{\partial X_{Si}} & \dfrac{\partial F_{2i}}{\partial Y_{Si}} & \dfrac{\partial F_{2i}}{\partial X_{Ti}} & \dfrac{\partial F_{2i}}{\partial Y_{Ti}} \end{vmatrix} \tag{E.34}$$

which can be evaluated as:

$$\mathbf{C}_i = \begin{pmatrix} a & b & -1 & 0 \\ -b & a & 0 & -1 \end{pmatrix} \tag{E.35}$$

The elements of the misclosure vector are determined from:

$$\mathbf{b}_i = -\begin{pmatrix} F_{1i}(\dot{\mathbf{x}},\mathring{\mathbf{s}}) \\ F_{2i}(\dot{\mathbf{x}},\mathring{\mathbf{s}}) \end{pmatrix} = -\begin{pmatrix} aX_{Si} + bY_{Si} + a_0 - X_{Ti} \\ -bX_{Si} + aY_{Si} + b_0 - Y_{Ti} \end{pmatrix} \tag{E.36}$$

The full set of equations now fits together as follows:

$$
\begin{pmatrix} \mathbf{A}_1 \\ \mathbf{A}_2 \\ \cdots \\ \mathbf{A}_n \end{pmatrix}
\begin{pmatrix} x_a \\ x_b \\ x_{a0} \\ x_{b0} \end{pmatrix}
+
\begin{pmatrix} \mathbf{C}_1 & 0 & 0 & \cdots & 0 \\ 0 & \mathbf{C}_2 & 0 & \cdots & 0 \\ \vdots & \vdots & \vdots & \ddots & \vdots \\ 0 & 0 & 0 & \cdots & \mathbf{C}_n \end{pmatrix}
\begin{pmatrix} v_{X_{S1}} \\ v_{Y_{S1}} \\ v_{X_{T1}} \\ v_{Y_{T1}} \\ v_{X_{S2}} \\ \vdots \\ v_{Y_{T2}} \end{pmatrix}
=
\begin{pmatrix} \mathbf{b}_1 \\ \mathbf{b}_2 \\ \vdots \\ \mathbf{b}_n \end{pmatrix}
\tag{E.37}
$$

The dimensions of these matrices are:

$$(2n \times 4)\,(4 \times 1) + (2n \times 4n)\,(4n \times 1) = (2n \times 1)$$

Note that each $\mathbf{0}$ in the $\mathbf{C}$ matrix is a sub-matrix of dimensions (2×4).

It is helpful to recall here that the equations derived above should be evaluated using the provisional values of the unobserved parameters (the transformation parameters) and the observed values of the observations (the coordinate sets). The equations could be made more explicit by writing terms such as $\dot{a}$, $\dot{b}$, $\Delta \dot{X}$, and $\Delta \dot{Y}$ for the transformation parameters, and $\dot{X}_{Si}$, $\dot{Y}_{Si}$, $\dot{X}_{Ti}$, and $\dot{Y}_{Ti}$ for the observed parameters; in practice, little confusion can arise as these will in general be the only values available.

Finally, to obtain a solution through the use of equation E.14, it is necessary to introduce an appropriate weight matrix. The most important point about this is that it must be consistent with the order of the observed parameters that has been implied by the formation of the $\mathbf{C}$ matrix and the $\mathbf{v}$ vector. Thus, the weight matrix $\mathbf{W}$ would take the form:

$$
\mathbf{W}^{-1} =
\begin{pmatrix}
\sigma^2_{XS1} & \sigma_{XS1YS1} & 0 & 0 & \sigma_{XS1XS2} & \cdots & \cdots \\
\sigma_{XS1YS1} & \sigma^2_{YS1} & 0 & 0 & \sigma_{YS1XS2} & \cdots & \cdots \\
0 & 0 & \sigma^2_{XT1} & \sigma_{XT1YT1} & 0 & \cdots & \cdots \\
0 & 0 & \sigma_{XT1YT1} & \sigma^2_{YT1} & 0 & \cdots & \cdots \\
\sigma_{XS1XS2} & \sigma_{YS1XS2} & 0 & 0 & \sigma^2_{XS2} & \cdots & \cdots \\
\vdots & \vdots & \vdots & \vdots & \vdots & \ddots & \vdots \\
\vdots & \vdots & \vdots & \vdots & \vdots & \cdots & \sigma^2_{YTn}
\end{pmatrix}
\tag{E.38}
$$

where σ^2_{XS1} is the variance of the observed coordinate of X_{S1}, and σ_{XS1YS1} is the covariance between coordinates X_{S1} and Y_{S1}, and so on.

This has been formed under the assumption that the two data sets are independent of each other, and thus that all the covariances between the two sets are zero.

E.2.2 The affine transformation

The affine transformation was introduced in section 4.5.4, the basic equations having the form:

$$X_T = a_0 + a_1 X_S + a_2 Y_S \tag{E.39}$$

$$Y_T = b_0 + b_1 X_S + b_2 Y_S \tag{E.40}$$

The functional relationships are therefore defined as:

$$F_{1i} = a_0 + a_1 X_{Si} + a_2 Y_{Si} - X_{Ti} \tag{E.41}$$

$$F_{2i} = b_0 + b_1 X_{Si} + b_2 Y_{Si} - Y_{Ti} \tag{E.42}$$

The sub-elements of the design matrices are:

$$\mathbf{A}_i = \left[\frac{\partial \mathbf{F}}{\partial \mathbf{x}} \right] = \begin{pmatrix} 1 & X_{Si} & Y_{Si} & 0 & 0 & 0 \\ 0 & 0 & 0 & 1 & X_{Si} & Y_{Si} \end{pmatrix} \tag{E.43}$$

$$\mathbf{C}_i = \left[\frac{\partial \mathbf{F}}{\partial \mathbf{s}} \right] = \begin{pmatrix} a_1 & a_2 & -1 & 0 \\ b_1 & b_2 & 0 & -1 \end{pmatrix} \tag{E.44}$$

And the misclosure vector is:

$$\mathbf{b}_i = -\begin{pmatrix} F_{1i}(\mathbf{\dot{x}}, \mathbf{\mathring{s}}) \\ F_{2i}(\mathbf{\dot{x}}, \mathbf{\mathring{s}}) \end{pmatrix} = -\begin{pmatrix} a_0 + a_1 X_{Si} + a_2 Y_{Si} - X_{Ti} \\ b_0 + b_1 X_{Si} + b_2 Y_{Si} - Y_{Ti} \end{pmatrix} \tag{E.45}$$

These elements fit together as:

$$\begin{pmatrix} \mathbf{A}_1 \\ \mathbf{A}_2 \\ \vdots \\ \mathbf{A}_n \end{pmatrix} \begin{pmatrix} x_a \\ x_a \\ x_a \\ x_{b_0} \\ x_{b_1} \\ x_{b_2} \end{pmatrix} + \begin{pmatrix} \mathbf{C}_1 & 0 & 0 & \cdots & 0 \\ 0 & \mathbf{C}_2 & 0 & \cdots & 0 \\ \vdots & \vdots & \vdots & \ddots & \vdots \\ 0 & 0 & 0 & \cdots & \mathbf{C}_n \end{pmatrix} \begin{pmatrix} v_{X_{S1}} \\ v_{Y_{S1}} \\ v_{X_{T1}} \\ v_{Y_{T1}} \\ v_{X_{S2}} \\ \vdots \\ v_{Y_{Tn}} \end{pmatrix} = \begin{pmatrix} \mathbf{b}_1 \\ \mathbf{b}_2 \\ \vdots \\ \mathbf{b}_n \end{pmatrix} \tag{E.46}$$

The dimensions of the matrices are:

$(2n \times 6)(6 \times 1) + (2n \times 4n)(4n \times 1) = (2n \times 1)$

Each $\mathbf{0}$ in $\mathbf{C}$ is again a (2×4) sub-matrix.
The weight matrix is again as in equation E.38.

E.2.3 Second order polynomials
The equations for the two dimensional transformation by second order polynomials were introduced in section 4.5.5 as:

$$X_T = a_0 + a_1 X_S + a_2 Y_S + a_3 X_S^2 + a_4 Y_S^2 + a_5 X_S Y_S \tag{E.47}$$

$$Y_T = b_0 + b_1 X_S + b_2 Y_S + b_3 X_S^2 + b_4 Y_S^2 + b_5 X_S Y_S \tag{E.48}$$

The functional relationships follow from these definitions:

$$F_{1i} = a_0 + a_1 X_{Si} + a_2 Y_{Si} + a_3 X_{Si}^2 + a_4 Y_{Si}^2 + a_5 X_{Si} Y_{Si} - X_{Ti} \tag{E.49}$$

$$F_{2i} = b_0 + b_1 X_{Si} + b_2 Y_{Si} + b_3 X_{Si}^2 + b_4 Y_{Si}^2 + b_5 X_{Si} Y_{Si} - Y_{Ti} \tag{E.50}$$

The sub-elements of the design matrices and the misclosure vector are:

$$\mathbf{A}_i = \left[\frac{\partial \mathbf{F}}{\partial \mathbf{x}}\right] = \begin{pmatrix} 1 & X_{Si} & Y_{Si} & X_{Si}^2 & Y_{Si}^2 & X_{Si}Y_{Si} & 0 & 0 & 0 & 0 & 0 & 0 \\ 0 & 0 & 0 & 0 & 0 & 0 & 1 & X_{Si} & Y_{Si} & X_{Si}^2 & Y_{Si}^2 & X_{Si}Y_{Si} \end{pmatrix} \tag{E.51}$$

$$\mathbf{C}_i = \left[\frac{\partial \mathbf{F}}{\partial \mathbf{s}}\right] = \begin{pmatrix} a_1 + 2a_3 X_{Si} + a_5 Y_{Si} & a_2 + 2a_4 Y_{Si} + a_5 X_{Si} & -1 & 0 \\ b_1 + 2b_3 X_{Si} + b_5 Y_{Si} & b_2 + 2b_4 Y_{Si} + b_5 X_{Si} & 0 & -1 \end{pmatrix} \tag{E.52}$$

$$\mathbf{b}_i = -\begin{pmatrix} F_{1i}(\dot{\mathbf{x}},\dot{\mathbf{s}}) \\ F_{2i}(\dot{\mathbf{x}},\dot{\mathbf{s}}) \end{pmatrix} = -\begin{pmatrix} a_0 + a_1 X_{Si} + a_2 Y_{Si} + a_3 X_{Si}^2 + a_4 Y_{Si}^2 + a_5 X_{Si}Y_{Si} - X_{Ti} \\ b_0 + b_1 X_{Si} + b_2 Y_{Si} + b_3 X_{Si}^2 + b_4 Y_{Si}^2 + b_5 X_{Si}Y_{Si} - X_{Ti} \end{pmatrix} \tag{E.53}$$

The full equations are then constructed as in equation E.46, except that the vector of unobserved parameters is:

$$\mathbf{x}^T = (x_{a0}\ x_{a1}\ x_{a2}\ \cdots\ x_{b4}\ x_{b5}) \tag{E.54}$$

The dimensions of the matrices are:

$$(2n \times 12)\ (12 \times 1) + (2n \times 4n)\ (4n \times 1) = (2n \times 1)$$

E.3 Three-dimensional transformations of Cartesian coordinates

E.3.1 The seven-parameter transformation

The seven-parameter three-dimensional geocentric transformation was introduced in section 4.3 in the form:

$$\begin{pmatrix} X \\ Y \\ Z \end{pmatrix}_T = (1 + \kappa)\ \mathbf{R} \begin{pmatrix} X \\ Y \\ Z \end{pmatrix}_S + \begin{pmatrix} \Delta X \\ \Delta Y \\ \Delta Z \end{pmatrix} \tag{E.55}$$

where $\mathbf{R}$ is the rotation matrix for the coordinate frame rotation convention given by:

$$\mathbf{R} = \begin{pmatrix} 1 & -\alpha_x & \alpha_y \\ \alpha_z & 1 & -\alpha_x \\ -\alpha_y & \alpha_x & 1 \end{pmatrix} \tag{E.56}$$

and α_x, α_y, and α_z are rotations about the X_S, Y_S, and Z_S axes, respectively. We have varied from the original form in replacing the scale factor μ (which in general is very close to unity) with the expression $(1 + \kappa)$, where κ is the small deviation from unity that is to be solved for directly.

The functional relationships follow from the above definition, and are conveniently expressed in matrix form as:

$$\mathbf{F}_i = (1 + \kappa)\ \mathbf{R}\mathbf{X}_S + \Delta\mathbf{X} - \mathbf{X}_T \tag{E.57}$$

where:

$$\Delta\mathbf{X} = \begin{pmatrix} \Delta X \\ \Delta Y \\ \Delta Z \end{pmatrix} \quad \mathbf{X}_T = \begin{pmatrix} X_T \\ Y_T \\ Z_T \end{pmatrix} \quad \mathbf{X}_S = \begin{pmatrix} X_S \\ Y_S \\ Z_S \end{pmatrix} \tag{E.58}$$

The design matrix **A** is then developed as follows:

$$\mathbf{A}_i = \left(\frac{\partial \mathbf{F}}{\partial \mathbf{x}}\right) = \left(\frac{\partial \mathbf{F}}{\partial \kappa} \quad \frac{\partial \mathbf{F}}{\partial \alpha_x} \quad \frac{\partial \mathbf{F}}{\partial \alpha_y} \quad \frac{\partial \mathbf{F}}{\partial \alpha_z} \quad \frac{\partial \mathbf{F}}{\partial \Delta \mathbf{X}}\right)$$

$$= \left(\mathbf{RX}_S \quad (1 + \kappa)\left(\frac{\partial \mathbf{R}}{\partial \alpha_x}\right)\mathbf{X}_S \quad (1 + \kappa)\left(\frac{\partial \mathbf{R}}{\partial \alpha_y}\right)\mathbf{X}_S \quad (1 + \kappa)\left(\frac{\partial \mathbf{R}}{\partial \alpha_z}\right)\mathbf{X}_S \quad \mathbf{I}\right) \text{ (E.59)}$$

The first element in equation E.59 can be evaluated as:

$$\mathbf{RX}_S = \begin{pmatrix} 1 & \alpha_z & -\alpha_y \\ -\alpha_z & 1 & \alpha_x \\ \alpha_y & -\alpha_x & 1 \end{pmatrix}\begin{pmatrix} X_S \\ Y_S \\ Z_S \end{pmatrix} = \begin{pmatrix} X_S + \alpha_z Y_S - \alpha_y Z_S \\ -\alpha_z X_S + Y_S + \alpha_x Z_S \\ \alpha_y X_S - \alpha_x Y_S + Z_S \end{pmatrix} \quad \text{(E.60)}$$

The most common application of this transformation is to the conversion of coordinates from one geodetic datum to another. In these situations, the rotations will be very small (a few seconds of arc) and it will be the case that:

$$X_S \gg a_z Y_S \quad \text{(E.61)}$$

and similar simplifications can therefore reduce equation E.60 to:

$$\mathbf{RX}_S = \begin{pmatrix} X_S \\ Y_S \\ Z_S \end{pmatrix} \quad \text{(E.62)}$$

The other terms in (E.59) are evaluated as:

$$(1 + \kappa)\left(\frac{\partial \mathbf{R}}{\partial \alpha_x}\right)\mathbf{X}_S = (1 + \kappa)\mathbf{R}\begin{pmatrix} 0 & 0 & 0 \\ 0 & 0 & 1 \\ 0 & -1 & 0 \end{pmatrix}\begin{pmatrix} X_S \\ Y_S \\ Z_S \end{pmatrix} = (1 + \kappa)\mathbf{R}\begin{pmatrix} 0 \\ Z_S \\ -Y_S \end{pmatrix}$$

$$= (1 + \kappa)\begin{pmatrix} \alpha_z Z_S + \alpha_y Y_S \\ Z_S - \alpha_x Y_S \\ -\alpha_x Z_S - Y_S \end{pmatrix} \approx (1 + \kappa)\begin{pmatrix} 0 \\ Z_S \\ -Y_S \end{pmatrix} \approx \begin{pmatrix} 0 \\ Z_S \\ -Y_S \end{pmatrix} \quad \text{(E.63)}$$

$$(1 + \kappa)\left(\frac{\partial \mathbf{R}}{\partial \alpha_y}\right)\mathbf{X}_S \approx \begin{pmatrix} -Z_S \\ 0 \\ X_S \end{pmatrix} \quad \text{(E.64)}$$

$$(1 + \kappa)\left(\frac{\partial \mathbf{R}}{\partial \alpha_z}\right)\mathbf{X}_S \approx \begin{pmatrix} Y_S \\ -X_S \\ 0 \end{pmatrix} \quad \text{(E.65)}$$

The complete set of equations forms the sub-element of the design matrix **A** thus:

$$\mathbf{A}_i = \begin{pmatrix} X_S & 0 & -Z_S & Y_S & 1 & 0 & 0 \\ Y_S & Z_S & 0 & -X_S & 0 & 1 & 0 \\ Z_S & -Y_S & X_S & 0 & 0 & 0 & 1 \end{pmatrix} \quad \text{(E.66)}$$

The sub-element of the design matrix **C** is formed from:

$$\mathbf{C}_i = \left[\frac{\partial \mathbf{F}}{\partial \mathbf{s}}\right] = ((1 + \kappa)\mathbf{R} \quad -\mathbf{I})$$

$$\approx \begin{pmatrix} 1 & \alpha_z & -\alpha_y & -1 & 0 & 0 \\ -\alpha_z & 1 & \alpha_z & 0 & -1 & 0 \\ \alpha_y & -\alpha_y & 1 & 0 & 0 & -1 \end{pmatrix} \quad \text{(E.67)}$$

The misclosure vector is given by:

$$\mathbf{b}_i = -\mathbf{F}_i\,(\dot{\mathbf{x}}, \mathring{\mathbf{s}}) = \Delta\mathbf{X} + (1 + \kappa)\mathbf{R}\mathbf{X}_S - \mathbf{X}_T \qquad \text{(E.68)}$$

The matrix sub-elements combine together in the full set of equations as:

$$
\begin{pmatrix} \mathbf{A}_1 \\ \mathbf{A}_2 \\ \vdots \\ \mathbf{A}_n \end{pmatrix}
\begin{pmatrix} x_\kappa \\ x_{\alpha_x} \\ x_{\alpha_y} \\ x_{\alpha_z} \\ x_{\Delta X} \\ x_{\Delta Y} \\ x_{\Delta Z} \end{pmatrix}
+
\begin{pmatrix} \mathbf{C}_1 & 0 & 0 & \cdots & 0 \\ 0 & \mathbf{C}_2 & 0 & \cdots & 0 \\ \vdots & \vdots & \vdots & \ddots & \vdots \\ 0 & 0 & 0 & \cdots & \mathbf{C}_n \end{pmatrix}
\begin{pmatrix} v_{X_{S1}} \\ v_{Y_{S1}} \\ v_{X_{T1}} \\ v_{Y_{T1}} \\ v_{X_{S2}} \\ \vdots \\ v_{Y_{Tn}} \end{pmatrix}
=
\begin{pmatrix} \mathbf{b}_1 \\ \mathbf{b}_2 \\ \vdots \\ \mathbf{b}_n \end{pmatrix}
\qquad \text{(E.69)}
$$

The dimensions of these matrices are:

$(3n \times 7)\,(7 \times 1) + (3n \times 6n)\,(6n \times 1) = (3n \times 1)$

E.3.2 The ten-parameter geocentric transformation

The ten-parameter geocentric transformation achieves a rotation about a local origin, $\mathbf{X}_0$, by re-formulating the transformation equations as:

$$
\begin{pmatrix} X \\ Y \\ Z \end{pmatrix}_T = (1 + \kappa)\,\mathbf{R} \begin{pmatrix} X_S - X_0 \\ Y_S - Y_0 \\ Z_S - Z_0 \end{pmatrix} + \begin{pmatrix} X_0 \\ Y_0 \\ Z_0 \end{pmatrix} + \begin{pmatrix} \Delta X \\ \Delta Y \\ \Delta Z \end{pmatrix}
\qquad \text{(E.70)}
$$

The local origin can be selected as the mean position of the source coordinates and therefore taken out of the least squares process. The derivation of the least squares formulation of this transformation then proceeds in a very similar way to the seven-parameter form. The functional relationship is altered to:

$$\mathbf{F}_i = (1 + \kappa)\mathbf{R}(\mathbf{X}_S - \mathbf{X}_0) + \mathbf{X}_0 + \Delta\mathbf{X} - \mathbf{X}_S \qquad \text{(E.71)}$$

The matrix $\mathbf{C}_i$ remains the same as before, but $\mathbf{A}_i$ becomes:

$$
\mathbf{A}_i = \begin{pmatrix}
X_S - X_0 & 0 & -Z_S + Z_0 & Y_S - Y_0 & 1 & 0 & 0 \\
Y_S - Y_0 & Z_S - Z_0 & 0 & -X_S + X_0 & 0 & 1 & 0 \\
Z_S - Z_0 & -Y_S + Y_0 & X_S - X_0 & 0 & 0 & 0 & 1
\end{pmatrix}
\qquad \text{(E.72)}
$$

The vector $\mathbf{b}$ is evaluated with the functional relationship in equation E.71, but all the matrices are combined as in equation E.69.

For either of these two models, the weight matrix is a three-dimensional version of that given in equation E.38 for the two-dimensional case.

E.3.3 Subsets of the seven-parameter geocentric transformation

In some situations, a full seven- or ten-parameter transformation is either not needed or is not possible with the existing data. It may be the case, for example, that only the average shift, or translation, between the datums is required. This would then be termed a three-parameter transformation, with only $(\Delta X, \Delta Y, \Delta Z)$ being determined.

To take another example, GPS data could be transformed into a local coordinate reference system whilst preserving the scale of the survey by applying a six-parameter transformation solving for $(\alpha_x, \alpha_y, \alpha_z, \Delta X, \Delta Y, \Delta Z)$.

Each of these is simply derived as a special case of the seven-parameter transformation, in which the columns of the matrix **A** that correspond to the unwanted parameters are eliminated. Thus, the design matrix for a solution of $(\Delta X, \Delta Y, \Delta Z)$ alone is found by eliminating the first four columns of equation E.66:

$$\mathbf{A}_i = \begin{pmatrix} 1 & 0 & 0 \\ 0 & 1 & 0 \\ 0 & 0 & 1 \end{pmatrix} \tag{E.73}$$

Similarly, the design matrix for a six-parameter transformation involving three shifts and three rotations is found by eliminating the first column of equation E.66:

$$\mathbf{A}_i = \begin{pmatrix} 0 & -Z_S & Y_S & 1 & 0 & 0 \\ Z_S & 0 & -Z_S & 0 & 1 & 0 \\ -Y_S & X_S & 0 & 0 & 0 & 1 \end{pmatrix} \tag{E.74}$$

In each case, the design matrix **C** will remain the same as in equation E.67.

E.4 Worked example

The way in which the solution is carried out is illustrated here with a worked example. The one selected is a three-dimensional seven-parameter transformation of geocentric coordinates. We have noted in section 4.5.3 that in many practical circumstances this method is not the best approach, but it is used here to illustrate the general principles for any transformation.

There are four points common to the two systems, and their coordinates are given as:

	System P		
Point	X	Y	Z
A	4027656.73	702.96	4973741.92
B	4025033.77	14050.08	4975857.89
C	4010282.95	1399.85	4987786.36
D	4009387.42	13295.68	4988482.31

	System Q		
Point	X	Y	Z
A	4027756.52	820.90	4973972.92
B	4025134.97	14168.85	4976087.46
C	4010381.77	1521.26	4988016.66
D	4009487.60	13417.55	4988711.44

The standard error of each coordinate in system P is 0.01 m, and in system Q is 0.02 m. For convenience, it will be assumed that all coordinates and points are uncorrelated, and that the weight matrix is therefore diagonal. It has the form:

$$\mathbf{W} = \begin{pmatrix} \mathbf{w} & 0 & 0 & 0 \\ 0 & \mathbf{w} & 0 & 0 \\ 0 & 0 & \mathbf{w} & 0 \\ 0 & 0 & 0 & \mathbf{w} \end{pmatrix}$$

where each sub-matrix **w** represents one of the points, and is given by:

$$\mathbf{w} = \begin{pmatrix} 1/0.01^2 & & & & & \\ & 1/0.01^2 & & & & \\ & & 1/0.01^2 & & & \\ & & & 1/0.02^2 & & \\ & & & & 1/0.02^2 & \\ & & & & & 1/0.02^2 \end{pmatrix}$$

As a first approximation, the provisional values of all the transformation parameters are set to zero, which will always be a valid assumption between any three-dimensional geocentric coordinate reference systems.

Each element of the design matrix **A** is then formed according to equation E.66, using the observed values of the coordinates. The full matrix is:

4027656.7	0.0	−4973741.9	703.0	1	0	0
703.0	4973741.9	0.0	−4027656.7	0	1	0
4973741.9	−703.0	4027656.7	0.0	0	0	1
4025033.8	0.0	−4975857.9	14050.1	1	0	0
14050.1	4975857.9	0.0	−4025033.8	0	1	0
4975857.9	−14050.1	4025033.8	0.0	0	0	1
4010282.9	0.0	−4987786.4	1399.9	1	0	0
1399.9	4987786.4	0.0	−4010282.9	0	1	0
4987786.4	−1399.9	4010282.9	0.0	0	0	1
4009387.4	0.0	−4988482.3	13295.7	1	0	0
13295.7	4988482.3	0.0	−4009387.4	0	1	0
4988482.3	−13295.7	4009387.4	0.0	0	0	1

The sub-elements of the design matrix **C** are as defined in equation E.67, which with all transformation parameters having provisional values of zero evaluates as:

$$\mathbf{C}_i = \begin{pmatrix} 1 & 0 & 0 & -1 & 0 & 0 \\ 0 & 1 & 0 & 0 & -1 & 0 \\ 0 & 0 & 1 & 0 & 0 & -1 \end{pmatrix}$$

These sub-matrices combine together into a full **C** matrix of dimensions 12×24, in the way shown in equation E.69.

The vector **b** is then evaluated from E.68, using the provisional values of the transformation parameters (all zero in the first iteration) and the observed values of the coordinates. Thus, the first sub-vector, b_1, is given as:

$$\begin{pmatrix} 99.79 \\ 117.94 \\ 231.00 \end{pmatrix} = \begin{pmatrix} 0 \\ 0 \\ 0 \end{pmatrix} + (1+0) \begin{pmatrix} 1 & 0 & 0 \\ 0 & 1 & 0 \\ 0 & 0 & 1 \end{pmatrix} \begin{pmatrix} 4027656.73 \\ 702.95 \\ 49733741.92 \end{pmatrix} - \begin{pmatrix} 4027756.52 \\ 820.90 \\ 4973972.92 \end{pmatrix}$$

And the full vector, including a similar computation for each point is:

$$\begin{matrix} 99.79 \\ 117.94 \\ 231.00 \\ 101.20 \\ 118.78 \\ 229.57 \\ 98.82 \\ 121.41 \\ 230.31 \\ 100.19 \\ 121.87 \\ 229.13 \end{matrix}$$

All necessary matrices have now been formed, and it remains only to carry out the matrix algebra solution of equation E.14 to determine the best estimates of the transformation parameter values. This gives the solution:

κ	2.0691E-05	= 20.69 ppm
α_x	9.90267E-05 rads	= 20.42"
α_y	4.99385E-05 rads	= 10.30"
α_z	0.00011886 rads	= 24.52"
ΔX	264.75	264.75 m
ΔY	104.14	104.13 m
ΔZ	−73.00	−73.00 m

The solution so obtained is added to the provisional set of values (in this case zero) to find the best estimates of the transformation parameter values. If the two datums of the coordinate reference systems were very different to each other, then in theory a further iteration would be required in which the current best estimates of the transformation parameter values become the provisional values. In most situations, however, this would not be required.

REFERENCES & FURTHER READING

Allan, A. L. (1997a) *Maths for Map Makers*, Whittles Publishing.

Allan, A. L. (1997b) *Practical Surveying and Computations*, Revised Second Edition, Laxtons.

Allan, A. L. (2007) *Principles of Geospatial Surveying*, Whittles Publishing.

Annoni, A., Luzet, C., Gubler, E., and Ihde, J. (2001) 'Map Projections for Europe', *European Union Joint Research Centre (JRC) Report EUR 20120 EN*. Available at: http://crs.bkg.bund.de/crs-eu

Ashkenazi, V., Crane, S. A., Preiss, W. J., and Williams, J. W. (1986) 'The 1980 Readjustment of the Triangulation of the United Kingdom and the Republic of Ireland OS(SN)80', *Ordnance Survey Professional Paper No. 31*.

BIPM (1983) *Resolution 1 of the 17th Conférence Générale des Poids et Mesures (CGPM)*, Bureau International des Poids et Mesures (BIPM). Available at: http://www.bipm.org

BKG (2007) *Information and Service System for European Coordinate Reference Systems*, Bundesamt für Kartographie und Geodäsie. Available at: http://crs.bkg.bund.de/crs-eu

Bomford, G. (1980) *Geodesy*, Fourth Edition, Clarendon Press, Oxford.

Bursa, M. (1966) 'Fundamentals of the Theory of Geometric Satellite Geodesy', *Geofysikalni Sbornik*, No. 241.

CNIG (1996) *Recommendations of the levelling network working group*, Conseil National de L'information Géographique (in French).

Cooper, M. A. R. (1987) *Control Surveys in Civil Engineering*, Collins.

Creel, T., Dorsey, A. J., Mendicki, P. J., Little, J., Mach, R. G., and Renfro, B. A. (2006) 'The Legacy Accuracy Improvement Initiative', *GPS World*, Mar 1.

Cross, P. A. (1983) 'Advanced Least Squares applied to Position Fixing', *Working Paper Number 6*, University of East London.

Dixon, K. (2007) 'Satellite Navigation', *Hydro International*, Vol. 11, No. 1, pp. 21–26.

FIG (2006) 'Guide on the Development of a Vertical Reference Surface for Hydrography', *International Federation of Surveyors Publication No. 37*.

Ghoddusi-Fard, R. and Dare, P. (2006) 'Online GPS processing services: an initial study', *GPS Solutions*, 10, pp. 12–20.

Hofmann-Wellenhof, B., Lichtenegger, H., and Collins, J. (1994) *GPS: Theory and Practice*, Third Revised Edition, Springer-Verlag.

Hooijberg, M. (1997) *Practical Geodesy using computers*, Springer.

ICSM (2002) *Geodetic Datum of Australia (GDA) Technical Manual*, Intergovernmental Committee on Surveying and Mapping. Available at: http://www.icsm.gov.au/gda

IHB (2005) *Manual on Hydrography, Publication M-13*, International Hydrographic Bureau. Available at: http://www.iho.shom.fr.

Iliffe, J. C., Ziebart, M., Cross, P. A., Forsberg, R., Strykowski, G., and Tscherning, C. C. (2003) 'OSGM02: A new model for converting GPS-derived heights to local height datums in Great Britain and Ireland', *Survey Review*, Vol. 37, No. 290, pp. 276–293.

Iliffe, J. C., Arthur, J. V., and Preston, C. (2007) 'The Snake Projection: a customised grid for rail projects', *Survey Review*, Vol. 39, No. 304, pp. 90–99.

IGS (2007) http://igscb.jpl.nasa.gov/

ISO (2007) *ISO 19111, Geographic information - Spatial referencing by coordinates*, Second Edition, International Organization for Standardization.

Johnstone, G. M. and Featherstone, W. E. 'AUSGeoid98: A new gravimetric geoid for Australia', *Australian Surveying and Land Information Group (AUSLIG) technical paper*. Available at: http://www.ga.gov.au/image_cache/GA5175.pdf

Junkins, D. R. and Farley, S. A. (1995) *NTv2 User's Guide*, Geomatics Canada. Available at: http://www.geod.nrcan.gc.ca/

Keay, J. (2000) *The Great Arc*, HarperCollins.

Leick, A. (2004) *GPS Satellite Surveying*, Third Edition, John Wiley and Sons.

Lemoine, F. G., (with 14 others) (1998) 'EGM96, The NASA Goddard Space Flight Center (GSFC) and National Imagery and Mapping Agency (NIMA) Joint Geopotential Model', *NASA Technical Paper 206861*. Available at: http://cddis.nasa.gov/926/egm96/egm96.html (NIMA is now known as the National Geospatial Intelligence Agency, NGA. See also NGA 2007.)

LINZ (2001) *Land Information Fact Sheet. New Zealand Transverse Mercator Projection*. Available at: http://www.linz.govt.nz/docs/miscellaneous/nztm.pdf

McCarthy, G. and Petit, G. (2003) 'IERS Conventions', *IERS Technical Note No. 32*. Available at: http://www.iers.org

Moritz, H. (1988) 'Geodetic Reference System 1980', *Bulletin Geodesique*, Vol. 62, No. 3. Available at: http://www.gfy.ku.dk/~iag/handbook/geodeti.htm

Mulcare, D. M. (2004a) 'The National Geodetic Survey NADCON Tool', *Professional Surveyor*, Vol. 24, No. 2.

Mulcare, D. M. (2004b) 'The National Geodetic Survey VERTCON Tool', *Professional Surveyor*, Vol. 24, No. 3.

NASA (1998) See Lemoine (1998).

NGA (1990) 'Datums, Ellipsoids, Grids, and Grid Reference Systems', *DMA Technical Manual 8358.1*. Available at: http://earth-info.nga.mil/GandG

NGA (2001) 'Department of Defense World Geodetic System 1984, Its Definition and Relationships With Local Geodetic Systems', *DMA Technical Report TR8350.2*, 3rd edition 1997, amended 2001. Available at: http://earth-info.nga.mil/GandG

NGA (2007) *NGA/NASA EGM96, N=M=360 Earth Gravitational Model*. Available at: http://earth-info.nga.mil:80/GandG/wgs84/gravitymod/egm96/egm96.html See also Lemoine (1998).

OGP (2007a) *The EPSG Geodetic Parameter Dataset*, International Association of Oil and Gas Producers. Available at: www.epsg.org

OGP (2007b) *Coordinate Conversions and Transformations including Formulas*, International Association of Oil and Gas Producers, Surveying and Positioning Guidance Note 7–2. Available at: www.epsg.org

Ordnance Survey (1999). *History of the Prime Meridian – Past and Present*. Available at: http://gpsinformation.net:80/main/greenwich.htm

Ordnance Survey (2006). 'A guide to coordinate systems in Great Britain', *Geodetic Information Paper No. 1*, Ordnance Survey of Great Britain, Southampton. Available at: http://www.ordnancesurvey.co.uk/

Ordnance Survey (2007). *Improved positioning using the National GPS Network – OSNet®*. Available at: www.ordnancesurvey.co.uk/osnet

OSI (2000) *The Irish Grid*, Joint Technical Information Paper, Ordnance Survey of Ireland, Dublin, and Ordnance Survey of Northern Ireland, Belfast.

POL (2007) *Chart Datum-Ordnance Datum: Differences at Selected UK Ports*, Proudman Oceanographic Laboratory, National Tidal and Sea Level Facility.

PROJ.4 *Cartographic projection programs*. Available at: http://kai.er.usgs.gov/ftp/index.html

Rapp, R. and Pavlis, N. (1990) 'The development and analysis of geopotential coefficient models to spherical harmonic degree 360', *Journal of Geophysical Research*, Vol. 95, No. B13, pp. 885–911.

Reilly, W. I. (1973) 'A conformal mapping projection with minimum scale error', *Survey Review*, Vol. 22, No. 168, pp. 57–71.

Roelse, A., Granger, H. W., and Graham, J. W. (1975) 'The adjustment of the Australian levelling survey 1970–71', *Division of National Mapping Technical Report No. 12*, 2nd Edition.

Russell, D. (2006) 'The impact of SATNAV developments on the DGNSS user and service provider', *Hydrographic Journal*, 119, 10–15.

Schmidt, K. (2000) 'The Danish height system DVR90', *Publication 4th Series, Vol. 8, National Survey and Cadastre*.

Sharma, S. K. (1966) 'A note on geodesic lengths', *Survey Review*, No. 140, pp. 291–295.

Snyder, J. P. (1981) 'Space Oblique Mercator Projection Mathematical Development', *Geological Survey Bulletin 1518*, Washington D. C., United States Government Printing Office.

Snyder, J. P. (1987) 'Map projections: a working manual', *Geological Survey Professional Paper 1395*, Washington D. C., United States Government Printing Office.

Stansell, T. A. (1978) *The Transit Navigation Satellite System*, Magnavox.

Stem, J. E. (1990) 'State Plane Coordinate System of 1983', *NOAA Manual NOS NGS 5*, Reprint with minor corrections. Available at: http://www.ngs.noaa.gov/PUBS_LIB/ManualNOSNGS5.pdf

Torge, W. (1991) *Geodesy*, Second Edition, Walter de Gruyter, Berlin.

UKHO (2007) *Admiralty Tide Tables, Volume 1*, United Kingdom Hydrographic Office, Taunton.

UNAVCO (2007) http://www.unavco.org

INDEX

3-parameter geocentric transformation 94, 155

7-parameter geocentric transformation 95–8, 157, 194

10-parameter geocentric transformation 98–100, 158, 196

Abridged Molodensky transformation method 102, 186

accuracy 18, 27, 31–3, 91–2
 loss due to transformation 91–2

adjustment of survey observations 24, 30

affine coordinate system 19, 35, 176

affine transformation 110, 115–17, 192

Albers equal area projection 75–6, 85

altitude 176

aspect 43, 63, 70

authalic projection
 see equal area projection 44, 51–3, 64, 85

azimuth
 astronomic 21
 geodetic 53, 182–4

azimuthal conformal projection 66–9, 71

azimuthal equal area projection 64, 85

azimuthal equidistant projection 63, 85

azimuthal orthographic projection 69–71

azimuthal perspective projection 70

azimuthal projections 42, 63–73, 85

bi-cubic interpolation 106

bi-linear interpolation 105, 126

Bursa-Wolf transformation 96

Cartesian coordinate system 4, 176

Cartesian coordinates
 for engineering applications 17
 geocentric 14, 178
 topocentric 18

central meridian
 see longitude of origin 49, 55

central meridian scale factor 57

change of coordinates 90–140

chart datum (CD) 33–4, 176

compound coordinate reference system (CCRS) 36, 176
 transformation of 130

concatenated transformations 106–8, 122, 127

conformal projection 53–6, 66–9, 75–81

conic conformal projection 45, 75–80

conic equal area projection 75

conic equidistant projection 75, 85

conic projection 42, 73–80

convergence 40, 55, 88

conversion 6, 91, 176
 accuracy 91
 ellipsoidal to geocentric Cartesian 15

coordinate 3, 176
 ellipsoidal 12–14
 geocentric Cartesian 14
 map projection 16

coordinate conversion 91, 176

coordinate frame transformation method 96, 157, 194

coordinate operation 176
 parameter 176
 parameter value 176

coordinate reference system (CRS)
    4, 7, 24–38, 177
  compound 36, 176
  definition of 36
  engineering 35, 178
  geocentric 25, 178
  geodetic 24–8, 178
  geographic 25, 179
  identification 36–8
  image 35, 179
  projected 28, 40, 180
  registers of 38
  vertical 29–36, 180
coordinate set 177
coordinate system 3, 11–20, 25, 177
  affine 19, 35, 176
  attributes 11
  cylindrical 19
  dimension 11
  ellipsoidal 12–14, 178
  linear 20
  spherical 14, 19
  topocentric 18
  vertical 18, 180
coordinate transformation 177
coordinate tuple 177
cylindrical conformal projection 53–6
cylindrical coordinate system 19
cylindrical equal area projection
    51–3, 85
cylindrical equidistant projection
    48–51, 85
cylindrical projections 42, 48–63

datum 4, 12, 20–35, 177–9
  definition 12, 21–3
  engineering 20, 35, 178
  geodetic 20–3, 178
  image 20, 35, 179
  realisation 24, 30
  vertical 20, 24–5, 29–36, 180
datum shift 177
datum transformation 177

depth 18, 177
developable surface 42–4
deviation of the vertical 21, 32
distance calculations 165
distances
  electronic 24
double stereographic projection 69

Earth
  centre of 22
  curvature 18
  model of 8–11, 20–3, 46–7, 177
  see also ellipsoid
easting 40, 49, 177
  at false origin 50
eccentricity 9
electronic distances 24
elevation 177
ellipsoid 8–11, 20–3, 46, 177
  computation of azimuth 184
  computation of coordinates 184
  defining parameters 9
  geometry of 14, 182
  radius 182
ellipsoidal coordinate system
    12–15, 178
ellipsoidal coordinates
  conversion to geocentric Cartesian
    coordinates 13, 15, 178
  transformation of 100–108
ellipsoidal height (h) 12–14,
    27–8, 32–3, 178
  transformation to gravity-related
    height 128–31
ellipsoidal latitude 178
ellipsoidal longitude 179
ellipsoidal normal 13, 32
engineering coordinate reference system
    35, 178
engineering datum 20, 35, 178
EPSG Geodetic Parameter Dataset
    10, 38, 133
equal area projection 44, 51–3, 64, 85

equidistant projection   44, 48–51,
     63, 74
errors   24−5, 30, 33
   measurement   25
ETRS89   26

false coordinates   39
false easting   50
false northing   50
false origin   50
flattening   9–10
foot   17
   international   17

Galileo
   see Global Navigation Satellite
      Systems (GNSS)   141–53
Gauss-Boaga   57
Gauss Conform   61
Gauss-Krüger   47, 57, 58
geocentric Cartesian coordinates
     14–19, 22, 24–6, 178
   conversion to ellipsoidal
      coordinates   15
   transformation of   23, 93–100
geocentric coordinate reference system
     25, 178
   transformations between   93–100
geocentric coordinates   178
geocentric translation transformation
     method   94, 155
geodesic   182, 185
geodetic azimuth   21, 53, 182–4
geodetic coordinate   178
geodetic coordinate reference system
     24–8, 178
geodetic coordinate system   178
geodetic datum   20–5, 27–8, 178
geodetic height   178
geodetic latitude   178
geodetic longitude   179
Geodetic Reference System of 1980
     (GRS 1980)   10, 22

geographic coordinate   178
geographic coordinate reference system
     25, 179
   transformations between   100–8
geographic offsets transformation
      method   102
geoid   8, 10–12, 18–23, 28–33, 179
   height   10, 31, 179
   model   32, 128
   undulation   31, 179
geoid-ellipsoid separation   20, 31,
     137, 179
geopotential number   19
Global Navigation Satellite Systems
     (GNSS)   7, 141–53
   augmentation systems   146
   differential   146
   kinematic   151
   positioning with codes   43
   positioning with phase
      observations   148
   systems   141
   transformation of coordinates
      120, 152, 154, 160
Global Positioning System (GPS)
     1, 7, 120, 161
   see also Global Navigation Satellite
      Systems (GNSS)   141–53
Glonass
   see Global Navigation Satellite
      Systems (GNSS)   141–53
gnomonic projection   69–70, 85
Google Earth™   72–3, 109, 118, 120,
     172–4
Google Maps™   46
graticule   12, 40–1
gravity-related height (H)   18, 32, 179
   transformation to ellipsoidal
      height   128–31
great circle   53
Greenwich prime meridian   22
grid   14, 28, 40–1, 169−71
   see map grid   40−1, 164

grid convergence
  see convergence   40, 55, 88
grid coordinates
  see Cartesian coordinates
  see also map projection coordinates
grid interpolation transformation
      method   104–6
grid origin   50

height
  ellipsoidal   12–13, 32, 178
  gravity-related   18, 179
height correction model   129
height offset   126, 128
  see also vertical offset   124, 126
Helmert transformation   96
Hotine Oblique Mercator projection   62

image coordinate reference system
      35, 179
image datum   20, 35, 179
international celestial reference sys-
      tem   22
International Terrestrial Reference
      Frame (ITRF)   25
International Terrestrial Reference
      System (ITRS)   22, 25
inverse flattening   9

Krovak projection   80

Lambert Conformal Conic (LCC)
      76–80, 164, 167
Laplace azimuth   21
latitude   4, 12–13, 27, 178
latitude of origin   49
least squares   187–98
linear coordinate system   20
local coordinate (reference)
      system   178
local datum   179
longitude   4, 12–13, 27, 179
  range   51

longitude of origin   49, 55
lowest astronomic tide (LAT)   33
loxodrome   53

map grid   40–1, 164
map projection   4, 39–89, 179
  area calculation   87, 166
  aspect   43, 63
  azimuthal   42, 63–73
  calculation within   86–9, 164–71
  conformality   53–6, 66–9, 75–81
  conic   42, 73–80
  convergence   40, 55, 88
  coordinates   16, 28
  cylindrical   42, 48–63
  designing   89, 162
  development of   46–8
  distance calculation   86, 165
  distortion   44–6, 49
  formula   83
  method   40–7
  non-geometric   80–3
  origin   39, 49
  orthomorphism
      see conformal projection
      53–6, 66–9, 75–81
  parameter values   40, 84–6
  re-scaling   39, 54
  reversibility   93
  shape distortion   168
  zone   40, 58–61, 76
mean sea level   8, 179
Mercator projection   46, 53–61
merging datasets   46
meridian   12, 40, 180
metre   17
  international   17
model of the Earth
  see ellipsoid   8–11, 20–3, 46, 177
Molodensky-Badekas transformation
      method   99, 158, 196
Molodensky transformation method
      102, 160–1, 186

NADCON   104
New Zealand Map Grid   81
normal height   19
normal section   182
northing   40, 49, 180–5
   at false origin   50
NTv2   104

oblique conic projection   80
Oblique Mercator projection   61–3
oblique stereographic projection   68
ordinate   3, 180
orthometric height   19
orthomorphic projection
   see conformal projection   45, 53–6,
      66–9, 75–81
OSTN02 transformation   106
overlay of dataset   118, 172

parallel (of latitude)   12, 40
perspective projection   70
Peters projection   52
plant north   35
plate carrée   49
polar projection   4, 63–8, 70
   see polar aspect   70
polynomial transformation
      110, 117, 193
position vector transformation
      method   95
prime meridian   12, 21, 180
   Greenwich   22
PROJ.4   69, 84
projected coordinate reference system
      28, 40, 180
projected coordinates   180
projection
   see map projection   39–89, 179
projection method
   see map projection method   40–7

Rectified Skew Ortomorphic
      projection   62

register of coordinate reference
      systems   38
   of transformations   132
rhumb line   53
rubber sheeting   117, 173

scale   9, 41
scale factor   40, 54, 86–8, 165–7
   at the (natural) origin   54, 57
semi-major axis   9
separation
   see geoid-ellipsoid separation
         20, 31, 137, 179
shape of the Earth   8–11
Similarity transformation method
         110, 112–16, 159, 189
similarity transformations   98
Snake projection   82
sounding datum   33
South-Oriented Transverse Mercator
      projection   61–2
Space Oblique Mercator projection
         62, 82
sphere   46
spherical coordinate system   14, 19
spherical coordinates
   computations with   181
spheroid   9, 177
standard parallel   54, 67, 73–4
stereographic projection method
      47, 66–9, 85
   see Universal Polar Stereographic
      (UPS)   67–8
survey observations
   adjustment of   24, 30

terminology   176–80
topocentric coordinate system   18
transformation   4, 90–140, 177
   accuracy   91–2, 152
   between ellipsoidal and gravity-related
      heights   128–31
   between geocentric CRSs   93–100

between geographic CRSs   100–8
deriving   134–40, 160, 187
indirect methods   106–8, 122, 127
multiplicity   91, 131
of compound CRSs   130
of ellipsoidal coordinates   100–8
of GPS coordinates   120, 152,
    154, 160
of plane coordinates   109–23, 182
of vertical coordinates   123–8
officially sanctioned   132, 160
registry   132
reversibility   93
selection of   131–4
transformation method
    Abridged Molodensky   101, 186
    affine   110, 115–17, 192
    Bursa-Wolf   96
    coordinate frame   96, 157, 194
    geocentric translations   94, 155
    geographic offsets   102
    geoid models   128
    grid interpolation   104–6
    height correction models   129
    Helmert   96
    Molodensky   101, 160, 186
    Molodensky-Badekas   99, 158, 196
    polynomial   110, 117, 193
    position vector   96
    similarity   97
    Similarity   110, 112–16, 159, 189

vertical offset and slope   126
vertical offsets   124
transformation parameter values
    determination by least squares   187
Transverse Mercator   47, 55–61

undulation of the geoid   31
Universal Polar Sterographic
    (UPS)   67–8
Universal Transverse Mercator (UTM)
    28, 58, 84

VERTCON   126
vertical coordinate reference system
    29–35, 180
    transformations between   123–8
vertical coordinate system   18, 180
vertical datum   20, 29–36, 180
vertical offset and slope transformation
    method   126
vertical offsets transformation
    method   124
Virtual Earth™   46

World Geodetic System of 1984
    (WGS 84)   10, 26, 151–3

zero meridian
    see prime meridian   12, 21, 180
zoning
    see map projection zone   40, 58–61, 76

Principles of Geospatial Surveying

ISBN: 978-1904445-21-0   234 × 156 mm   472pp   illustrated   softback   £65

Dr Arthur L. Allan, Emeritus Reader in Surveying, University College London, UK

> ➤ The handbook for a thorough understanding of the principles of modern geospatial surveying techniques

This important new book replaces the author's highly successful Practical Surveying and Computations and has been completely recast to accord with modern practices of geospatial surveying.

Since most practical work is carried out by prescribed systems and processed by software packages, the book concentrates on those essential principles which the user needs to know, if the results are to be verified and assessed with understanding and wisdom. The text outlines the fundamentals of geospatial surveying including relevant worked examples that make liberal use of Excel spreadsheets. The mathematical treatment relates directly to those topics found in the author's successful textbook, Maths for Map Makers.

Contents: Introduction to geospatial surveying; Technical procedures; Coordinate systems; Coordinate transformations; Theory of errors and quality control; Least squares estimation; Satellite surveying; Survey computations; Heights and levels; Maps and map data processing; Construction and curves; Industrial and engineering surveying; Instrumentation. Appendices: Useful data; Spherical trigonometry; General least squares; The Earth ellipsoid; Survey control for photogrammetric mapping; Quality control; Field astronomy; Survey projections; Satellite surveying; References

Readership: This book will fill the gap between elementary and advanced texts for students and professionals working in the fields of land and engineering surveying, applied geomatics, civil and industrial engineering and other geospatial applications such as building and structural monitoring, road accident appraisal and field sports adjudication.

Maths for Map Makers, second edition

ISBN 1-870325-99-0   234 × 156mm   softback   liberally illustrated   416pp   £29.99

Dr Arthur L. Allan, Emeritus Reader, University College London

This book remains the first port of call for anyone seeking a comprehensive treatment of the mathematics involved in map making. In response to demand, the author has written a new chapter on the subject of least squares estimation. To support this new topic, an appendix has been added, presenting additional matrix algebra. The new text is again problem-oriented, and is generously illustrated with worked examples and exercises.

Contents: Numbers and calculation; Plane geometry; Trigonometry; Plane coordinates; Problems in three dimensions; Areas and volumes; Matrices; Vectors; Calculus; Conic sections; Spherical trigonometry; Solution of equations; Leastsquares estimation; References. Useful data. Further matrix algebra. Notation for least squares. The error ellipse and its pedal curve. Summary of formulae. Index

This is an ideal course book for students of geomatics including surveying, cartography, photogrammetry, geography and civil engineering and anyone in industry who requires a reference work on the topic

'… This book continues to be an excellent text for students of geomatics and associated subjects. … an important aid to revision and enhanced understanding'. *The Photogrammetric Record*

'… This is a new edition and further endorses the work as the first port of call for anyone seeking a comprehensive treatment of the mathematics involved in map making and elementary surveying. … This is an ideal course book for students of geomatics including surveying, cartography and Photogrammetry …' *Geomatics World*

Whittles Publishing, Dunbeath, Caithness, Scotland, KW6 6EY, UK
T: +44(0)1596-731 333; F: +44(0)1593-731 400; E: info@whittlespublishing.com
www.whittlespublishing.com